A121

W9-AQJ-929

URBAN
ECONOMIC
DEVELOPMENT

Volume 27, URBAN AFFAIRS ANNUAL REVIEWS

INTERNATIONAL EDITORIAL ADVISORY BOARD

ROBERT R. ALFORD, *University of California, Santa Cruz*
HOWARD S. BECKER, *Northwestern University*
BRIAN J.L. BERRY, *Carnegie-Mellon University*
ASA BRIGGS, *Worcester College, Oxford University*
JOHN W. DYCKMAN, *Johns Hopkins University*
SPERIDIAO FAISSOL, *Rio de Janiero State University*
JEAN GOTTMANN, *Oxford University*
SCOTT GREER, *University of Wisconsin—Milwaukee*
BERTRAM M. GROSS, *Hunter College, City University of New York*
PETER HALL, *University of Reading, England*
ROBERT J. HAVIGHURST, *University of Chicago*
EHCHI ISOMURA, *Tokyo University*
ELISABETH LICHTENBERGER, *University of Vienna*
M. I. LOGAN, *Monash University*
WILLIAM C. LORING, *Center for Disease Control, Atlanta*
AKIN L. MABOGUNJE, *University of Ibadan*
MARTIN MEYERSON, *University of Pennsylvania*
EDUARDO NEIRA-ALVA, *CEPAL, Mexico City*
ELINOR OSTROM, *Indiana University*
P.J.O. SELF, *London School of Economics and Political Science*
WILBUR R. THOMPSON, *Wayne State University and Northwestern University*

URBAN ECONOMIC DEVELOPMENT

Edited by

RICHARD D. BINGHAM
and
JOHN P. BLAIR

Published in cooperation with the Urban Research Center, University of Wisconsin—Milwaukee

Volume 27, URBAN AFFAIRS ANNUAL REVIEWS

SAGE PUBLICATIONS
Beverly Hills   London   New Delhi

Copyright © 1984 by Sage Publications, Inc.

All rights reserved. No part of this book may be reproduced or utilized in any form or by any means, electronic or mechanical, including photocopying, recording, or by any information storage and retrieval system, without permission in writing from the publisher.

For information address:

SAGE Publications, Inc.
275 South Beverly Drive
Beverly Hills, California 90212

SAGE Publications India Pvt. Ltd.
C-236 Defence Colony
New Delhi 110 024, India

SAGE Publications Ltd
28 Banner Street
London EC1Y 8QE, England

Printed in the United States of America

Library of Congress Cataloging in Publication Data

Main entry under title:

Urban economic development.

(Urban affairs annual reviews ; v. 27)
"Published in cooperation with the Urban Research
Center, University of Wisconsin—Milwaukee."
1. Community development, Urban—United States—
Addresses, essays, lectures. 2. Urban economics—
Addresses, essays, lectures. 3. United States—Economic
conditions—1981-   —Addresses, essays, lectures.
I. Bingham, Richard D. II. Blair, John P., 1947-
III. University of Wisconsin—Milwaukee. Urban Research
Center. IV. Series.
RT108.U7   vol. 27   307.7′6 s   [338.973]   84-11456
[HN90.C6]
ISBN 0-8039-1998-0
ISBN 0-8039-1999-9 (pbk.)

FIRST PRINTING

Contents

Preface

☐ URBAN ECONOMIC DEVELOPMENT is the second volume of the Urban Affairs Annual Reviews cosponsored by Sage Publications, Inc. and the Urban Research Center of the University of Wisconsin—Milwaukee. Our task at the Center was to identify significant issues of concern to our urban society and to select well-qualified authors to design and edit volumes concerned with these issues. With the present volume, we believe that we have been successful in both regards.

The economic decentralization implicit in President Reagan's proposal to increasingly rely on state and local government, in cooperation with the private sector, to restore economic growth is a major shift in American federalism. Throughout our recent history there has been a history of tension between those advocating increased national responsibilities and those favoring state responsibilities and local initiatives. From the inception of the New Deal through the latter half of the Carter administration years, a substantial amount of decision making had been transferred to the national government. When President Reagan took office, however, he reversed this trend—declaring his intention "to curb the size and influence of the federal establishment and to demand recognition of the distinction between the powers granted to the federal government and those reserved to the states or to the people." This view presumes that the federal government can change the nation by withdrawing from many of its traditional activities.

When the Reagan administration took office it inherited a range of geographically targeted federal economic development programs, housed primarily in three agencies—the Economic Development Administration (EDA), the Appalachian Regional Commission (ARC), and the Department of Housing and Urban Development (HUD). The EDA and ARC administered a variety of grants, loans, and loan guarantees for community and economic development projects, infrastructure improvements, and, in some cases, direct assistance to firms. HUD administered the Urban Development Action

7

Grant (UDAG) program, which provided competitive economic development grants to local government. The administration proposed to eliminate all funding for the EDA and ARC and to substantially reduce UDAG. These programs were to be replaced by a new "supply-side" proposal for urban enterprise zones. While the President was not successful in all his initiatives, he was successful in substantially reducing the overall funding of federal economic development assistance. In addition, a number of states have now initiated their own enterprise zone programs.

The depth of the recession of the 1980s, the change in the economic structure of the nation, and the dramatic shift in our federal economic development efforts have all combined to make economic development one of the most important urban issues of the decade. Bingham and Blair have done much to help us understand these issues.

University of Wisconsin-Milwaukee   *Ann Lennarson Greer*
                                                      Scott Greer

Acknowledgments

☐ WE WISH to thank the staff of the Urban Research Center of the University of Wisconisn-Milwaukee and the Department of Economics at Wright State University for their many contributions to this book. Wanda Wimes, Kathy Lelinski, and Penny Stacy spent many hours typing chapters, correcting errors, and coordinating the work of the authors. Laura Moser provided valuable editorial assistance and pointed out many areas that needed clarification and improvement. We are grateful for their efforts.

John P. Blair
Richard D. Bingham

Introduction:
Urban Economic Development

RICHARD D. BINGHAM
JOHN P. BLAIR

☐ URBAN ECONOMIC DEVELOPMENT is one of the hottest topics of the decade. A recent description of the attention given to economic development among planners applies to other urban professionals as well (Bergman, 1983: 260): "Acceptance came so rapidly and completely that long-time planners now overlook the fact that very few of them posed it as an important planning activity a decade ago."

While economic development has always been important to the nation and its cities, it was not until the late 1970s and early 1980s that explicit recognition was given to the need for the public and private sectors to work together to expand the economic base of central cities and metropolises. Prior to this time, the public and private sectors were frequently in conflict. States found it easier to raise corporate taxes rather than taxing citizens directly, and governors could call on business to "pay their fair share." Urban officials often viewed economic development projects as competing with poverty programs for funding. Many communities and some states had well-publicized antigrowth philosophies. Today the situation is entirely different. Governors, legislators, and mayors cringe when they hear accusations that their jurisdiction is "unfriendly to business" or that the "business climate in the state is poor." Economic development is now seen as an avenue for ameliorating problems of poverty (although it does not have the negative connotation often associated with poverty programs) and increasing urban fiscal capacity.

This is not to suggest that state and local governments have historically ignored private development—they have not. Every state in the United States administers an economic development program offering incentives to private industry. However, the widespread legitimacy and institutionalization of ongoing bureaucratic-based programs has emerged in the last decade. The practice of providing financial incentives to business is rooted more firmly in state government than at any other level in the federal system, although cities and counties are rapidly developing their own programs and agencies to implement them.

Part of the reason for the interest in urban economic development can be traced to the Reagan administration's concern about national economic growth. Early in Reagan's term, attention to macroeconomic affairs threatened to swamp concern for poverty and life quality urban programs. Officials within the Department of Housing and Urban Development were in step with the President. The *President's National Urban Policy Report, 1982* (HUD, 1983: 13) stated that "stabilizing and revitalizing the *national* economy is the most important Federal urban policy for the 1980's" (italics added). The administration's focus on national economic growth was coupled with beliefs that undermined traditional approaches to urban problems. Many federal government activities were considered a drag on private sector activity, and programs designed to help or protect the urban poor were particularly cited as impediments to aggregate economic growth. Cutting social programs to reduce deficits and stimulate work as well as to eliminate cost-ineffective regulations were part of the supply-side economic recovery strategy.

However, the administration did not choose to advocate a classic laissez faire approach to urban social and economic problems. Federal urban programs were cut back, but if the administration desired a free market approach, it could have at least used moral suasion to discourage state and local involvements in business affairs. Instead, conservative idealogy was subordinated to the desire to decentralize economic development activity. Thus, the administration actually encouraged state and local governments to become actively involved in stimulating private business activities to solve urban problems. The Task Force on Private Sector Initiatives was formed to develop ways that public private partnerships could encourage economic activity. The administration pointed to the need to cooperate with private business interests and leveraging of private dollars has become an important tactic (HUD, 1982). Although many observers believe that the call for more direct local involvement with private business was a smokescreen for reducing federal activities and expenditures, the policy tactic of public-private partnerships led to an increase in public subsidies (or incentives depending on viewpoint) to encourage local business activity.

The most important economic development incentives include tax reductions, below market interest rate financing, and publically financed infrastructure. As public-private partnerships become routine, urban administrators are becoming as comfortable with business concepts such as balloon payment, equity, and accelerated depreciation as they have been with community development terms such as "citizen participation" and "low-mod benefits."

Although the conservative desire to stimulate aggregate economic growth was the most pronounced intellectual influence on the emergence of urban economic development, liberals have also been concerned about sluggish macro-economic performance and saw an urban role in revitalization. Many liberal economists and urbanists were arguing for sweeping changes in economic institutions and values because, like conservatives, they were disturbed by the Carter administration's inability to manage the economy. For instance, Theobald (1983: 1) contended that

> economists have become expert in coining jargon to disguise their inability to understand what is going on . . . . What is needed is imaginative thinking about ways of restructuring lifestyles and lifecycles . . . and . . . ways of ensuring security, developing skills, distributing resources, and providing prestige.

As intriguing as "radical restructuring" appeals may be, urban development officials have almost universally remained within the institutional and value framework of traditional capitalism. The creativity of local development officials has been substantial. But the creativity has been directed at developing new (and complicated) arrangements between governments, businesses, and third sector organizations. The variety of subsidy/compensation schemes is enormous. However, the idea that a better city can be achieved by more jobs, money, tax revenue, and so forth is entrenched in local economic development efforts.

The current "pro-business" philosophy was recently summarized by the Urban Institute (1983: 4):

> It is now widely recognized that the pubic sector often unintentionally influences private business. Recently, public tax and regulatory policies have been cited as serious forces in the private decision-making process. If these policies have unintentionally discouraged business growth, and if, as has been alleged, federal highway and housing programs helped fuel the exodus from inner cities, then it is not unreasonable to expect that government can intentionally influence private sector actions.

Similarly, the Urban Policy Report proclaimed (HUD, 1982: 23):

> In general, Federal economic development programs were created to increase private sector investment in communities experiencing economic decline . . . . These programs had the unintended effect of channelling credit to less competitive firms.

> The private market is more efficient than Federal program administrators in allocating dollars among alternative uses. In this period when some communities are experiencing job losses and contracting tax bases . . . federal assistance is intended to complement rather than displace market decision process . . . . State and local governments have primary investors. They are most likely to succeed if they form partnerships with these private sectors and plan strategically to enhance their comparative advantage relative to other jurisdictions.

Private sector representatives blame the public sector for many of their problems and this attitude has affected the nature of current economic development programs. It contributed to the emphasis on public/private cooperation or partnerships. In a recent report to the president, the Business-Higher Education Forum (1983: 24) held that:

> In large part, the decline of U.S. economic competitiveness is a result of fragmented, unfocused and changing public policy-making processes. This "ad hocism" is important because government actions create the fundamental environment in which the private sector conducts its business.

> At the macro level, fiscal and monetary policies shift frequently and abruptly—creating uncertainty, instability, and recurring boomlets and declines. At the micro level, these same *ad hoc* processes have characterized the development and implementation of a wide range of policies, including those on trade and investment, regulation, antitrust, development of human resources, urban and rural development, public works, natural resources, research and development, and strategic materials. The design and administration of these policies virtually guarantee that their limited objectives will be pursued with little regard for broader national goals or for their effects on other policy objectives.

> Not surprisingly, policies frequently operate at cross purposes and undermine each other. For example, the virtual explosion of regulatory activities in the 1970s annually diverted tens of billions of private investment capital into non-productive activities at the same time that fiscal policies were attempting to stimulate productive capital investment in modernized plant, equipment and technology.

The view that government was ineffectual or perverse in affecting the economy ignores successes—reducing poverty, accommodating the entrance of the baby boom into the work force and improving environmental standards (Schwarz, 1983). It nevertheless reflects an opinion that contributed toward the marriage of public economic development initiatives and business interests.

A related reason for the merging interest in local economic development is the distressed conditions of many urban areas. A decade ago, public concern was focused on the economic condition of minorities and other disadvantaged groups. Economic development was one of several ways of helping them, but it was not the dominant approach. Today the shift from manufacturing to service-based economies coupled with slow overall growth has greatly broadened the constituency group for local economic development efforts. Many dislocated blue-collar workers are members of influential unions or neighborhood constituency groups. Even workers who have jobs in high demand occupations often feel threatened by rapid change that could force them to relocate if they want work. Economic shifts are altering the locational advantages of regions and high anxiety about economic futures is widespread. Although some observers have suggested that the growing service sector will take up the employment slack, many manufacturing workers will lack the ability to retrain. Furthermore, Noyelle (1983) showed that recent urban transitions are uneven and lower ranking urban areas that are weak in advanced services face enormous challenges.

The federal government is also reducing its financial role in attempting to alleviate the conditions of distressed areas, thus challenging state and local officials to be innovative in economic development activities. The major remaining federal program, Urban Development Action Grants (UDAG), requires a high level of local government and business/industry cooperation and major contributions of private capital to projects. Furthermore, if the present administration has its way, UDAG will be further reduced and an Urban Enterprise Zone program requiring no direct financial contribution by the federal government will become the flagship urban program. State and local governments have been challenged to fill the vital role of financing urban economic development.

Although the current concern about economic development sprang from a conservative urban policy, the process of intensive state and local participation in private business activities is inconsistent with traditional conservative principles. If the trends described in this volume are extrapolated, it is easy to envision an economic society that includes "corporate cities"—cities that have a substantial equity investment in local businesses and receive equity returns. The city investments may have been by direct or indirect subsidy through tax forgiveness and infrastructure provision. As a result of their contributions, urban governments may gain some control over local business decisions. There are isolated examples of cities owning partnership interests in real estate developments. Current urban development programs represent a potential source of quasiprivate enterprise that could affect the nature of major economic institutions. This development could be an ironic twist for a conservatively inspired urban policy.

Finally, we have entered a period of heightened interstate and interlocal competition for new private investments. State and local governments today are being forced to examine their functions and attitudes. Public infrastructure is deteriorating rapidly at just the time governments are being forced to cut back on expenditures (Buss and Redburn, 1983).

THE PLAN OF THE BOOK

The purpose of this book is to explore the issues surrounding urban economic development and to see how urban economic development actually works. The chapters that follow cover four major areas: economic development paths, federal economic development programs, state and local approaches to economic development, and local economic development at work.

ECONOMIC DEVELOPMENT PATHS

Unfortunately, much that the Business-Higher Education Forum says about the ad hoc and reactive nature of economic development strategies is true—especially at the state and local levels. This section seeks to place urban economic development in perspective. In introducing the sections that follow, it is necessary to step back and look at the grand scheme of things. The four chapters in the section are designed to provide this background. We begin with a historical view of economic development in the United States. Benjamin describes some historical approaches to urban economic development. We then look at some specific economic development paths that public entities can adopt. Premus describes opportunities to communities that can attract firms on the frontiers of technology. Knight and Gappert show that some cities can prosper by becoming international centers for corporate activity. Whatever path is used for development, entrepreneurship will be an important ingredient. Thus Chen reviews determinants of entrepreneurship among minorities and provides some policy suggestions. There are obviously many more paths that could be explored if space were available. Examples include central business district revitalization, government centered development, and development as an artistic center. The readings are designed to stimulate the reader into thinking about these broader strategic issues.

FEDERAL ECONOMIC DEVELOPMENT PROGRAMS: ANALYSIS AND EVALUATION

There is presently a significant debate going on about the "best" way or ways in which the federal government can stimulate economic development in distressed areas. The evaluation research literature does not point to a definitive single approach. Two of the three chapters in this section evaluate the economic development potential of two types of programs: block grants and categorical grants. The Community Development Block Grant Program described by Dommel provides local policymakers with discretion not only about what type of economic development projects they select but whether economic development or some other activity should be undertaken. The Urban Development Action Grant Program requires cities and businesses to work together to develop a proposal that must compete with other proposals for federal funding. Gatons and Brinthall describe and evaluate this high profile federal program. The third chapter in this section examines the concept of enterprise zones—the president's major urban initiative now before Congress. Butler, one of the foremost proponents of the idea, provides an overview of the concept and rationale.

The section is intended to give readers a general background in terms of the types of programs designed to aid specific targeted distressed areas and the particular philosophies they represent. The programs presented by no means cover all federal economic development programs because good summaries exist elsewhere (National Council for Urban Economic Development, 1982). This section, however, represents the core development programs of the federal government that will probably be the bases of future programmatic changes.

STATE AND LOCAL APPROACHES TO ECONOMIC DEVELOPMENT

One of the widely discussed advantages of a federal system is that alternative administrative techniques can be tried and, if successful, they can be copied by other local governments. If unsuccessful, the cost will be relatively low. However, the federal system also leads to competition that can be destructive. The wisdom of state and local competition was widely debated when Volkswagen openly encouraged communities to bid against each other to be the site of their production facilities (Gorden and Lees, 1981). In the past several years, the competition has stiffened. For instance, there are currently about two thousand firms attempting to aid community economic development efforts. Over $100,000,000 is spent annually on advertizing alone.

Two chapters in this section describe issues of cooperation and competition among local governments in economic development efforts. James shows that the rationale for urban economic development programs needs bolstering and describes the kinds of evidence that is needed. Neenan and Ethridge give an account of intergovernmental competition with the framework of the classic Tiebout model of residential choice. They show that there are both positive and negative aspects of interurban competition for jobs. The third chapter, by Litvak, describes how cities are using pension funds, an increasingly important resource, to finance economic development.

LOCAL ECONOMIC DEVELOPMENT AT WORK

The final section of the book is a real eye-opener. It contains four case studies of local economic development at work. Two of the studies are of cities. Berkowitz describes the relatively successful case of Baltimore, and Felbinger analyzes the relatively unsuccessful revitalization efforts of Joliet. The other two cases are of city-corporate relationships. Jones and Bachelor describe Detroit's attempt to secure a General Motors facility as the "Politics of Production." Blair and Wechsler present a similar situation (in which Fort Wayne and Springfield compete to avoid having their International Harvester plant closed) from an economic choice perspective. The questions raised by this section are important. Can local economic development programs "turn around" dying cities? If so, what programs work and at what costs? Can cities bargain successfully with large corporations and are some cities in too weak a position to engage in interurban competition for jobs?

REMARKS

This volume is a thought-provoking introduction to the concept of economic development. It is intended to stimulate strategic thinking about solutions to a complex web of problems that affect the lives and livelihood of urban Americans. We are fortunate that many of the authors are nationally known experts on these issues and we thank them for their contributions. We hope the readings that follow will stimulate further consideration of the emerging relationships between the public and private sectors and how these relationships will affect urban economic life.

REFERENCES

BERGMAN, E. M. (1983) "Introduction." Journal of the American Planning Association 49, 3 (Summer): 260-262.

Business-Higher Education Forum (1981) America's Competitive Challenge: The Need for a National Response. Washington, DC: Business-Higher Education Forum.

BUSS, T. F. and F. S. REDBURN (1983) "Introduction essay." Southern Review of Public Administration 6 (Winter): 384-389.

Committee for Economic Development (1982) Public-Private Partnership: An Opportunity for Urban Communities. Washington, DC: Committee for Economic Development.

GORDON, S. L. F. A. LEES (1981) "Locating foreign plants in the U.S.: the Volkswagen experience." Industrial Development 150, 6 (Nov.-Dec.): 24-27.

National Council for Urban Economic Development (1982) CUED's Guide to Federal Economic Development Programs. Washington, DC: National Council for Urban Economic Development.

NOYELLE, T. J. (1983) "The rise of advanced services." Journal of the American Planning Association 49, 3 (Summer): 280-290.

SCHWARZ, J. E. (1983) America's Hidden Success. New York: W. W. Norton.

THEOBALD, R. (1983) "Work and leisure, employment and unemployment: struggling towards new perceptions." CUE Bulletin 10, 12 (May): 1-4.

Urban Institute (1983) Director of Incentives for Business Investment and Development in the United States. Washington, DC: Urban Institute Press.

U.S. Department of Housing and Urban Development (1982) The President's National Urban Policy Report, 1982. Washington, DC: Department of Housing and Urban Development.

Part I

Economic Development Paths

1

From Waterways to Waterfronts: Public Investment for Cities, 1815-1980

ROBERT L. COOK BENJAMIN

☐ URBAN HISTORY and urban economic development are relatively new academic disciplines in the United States. Despite some useful monographs, their common ground of a history of American urban economic development is unformed. In order to shape such a history, this chapter focuses upon government investment in urban economic development from 1815 onward. The causes, types, magnitude, and distribution of such government investment will provide the major subthemes.

The periods chosen to arrange the theme and subthemes chronologically—The Canal and Railroad Race of 1815 to 1870, The Great City Era of 1870 to 1929, The Depression Decade of 1929 to 1939, The Urban Renewal Era of 1940 to 1965, The Retreat from Economic Development of 1966 to 1974 (covered lightly), and Making the Most of Scarcity of 1975 to 1980—have some of the arbitrariness of all historical periods. But they express well the different phases of government involvement in urban economic development, and they also capture some of the major shifts in city economic status.

Despite the definite role of human investment in national and local economic growth, physical investments that enhanced property values and created jobs will be the focus of economic development in this overview. Such investments were usually what governments meant by economic development (though, in practice, various levels of government made impressive contributions to economic growth through human investments such as public colleges).

THE CANAL AND RAILROAD RACE
OF 1815-1870

Urban economies and population took off in the United States after 1815 as canals, turnpikes, and railroads developed internal resources and markets and as faster sailing ships and steamships expanded European markets and migration. Until the transportation revolution, the American urban economy had been limited by forbidding costs of shipment over land. In 1816, for instance, it was cheaper to send European goods to Pittsburgh by way of New Orleans and its river routes than by way of Philadelphia.

The prize of being the first to tap an interior market was great. It first inspired an era of competition between seaboard cities/states against other seaboard cities/states. The phenomenal success of the Erie Canal for New York City and State redoubled the transportation efforts of other cities and states, first with canals and then with railroads. Soon after, so rapid was the development of the interior, hundreds of emerging towns vied within each state for locational preference as canal and then as railroad termini (or as county seats and state capitols). The towns that successfully lobbied the state legislature for these advantages enjoyed an economic multiplier of commerce, residents, and further transport linkages that established local economic dominance (Chudacoff, 1981: 38-40). Before 1860, many towns and cities also petitioned Congress directly for transportation assistance and were often successful.

Established cities in the 1820s to 1840s and then some of the larger interior cities of the 1840s to 1860s directly financed transportation routes, especially railroads, in a bid to tap markets. Most notably, Baltimore aided its merchants in financing America's first commercial railroad when direct canal routes to the West were stalled by natural and legislative obstacles (Olson, 1980: 71-77). In the aggregate, however, states provided the greater amount of tax, land, and credit resources for transportation development.

The transportation revolution in the United States of 1815 to 1860 was much more a public undertaking than the concurrent and earlier transportation revolutions in England. Relative to England, the United States had much less experience with corporate forms and had much less private capital. (Indeed, British investors bought many of the transportation bonds issued by American local governments.) Equally important, the American economic landscape offered many fewer established and closely settled areas of trade than England's. American transportation investment had to "depend for most of its prospective revenue on the settlement and economic activity which its own construction was intended to promote"—a classic condition for public support of development (Goodrich, 1960).

State and city rivalry to build a transportation network led to some overbuilding and to state and city defaults during the financial crisis of the late 1830s and early 1840s. But, overall, the investments were exceptionally productive for all levels of the economy. State and local incentives to build railroads after 1850 were overshadowed by federal incentives and route choices, for which states and cities competed fiercely.

As the transportation revolution intensified, the United States in the 1840s and 1850s had the most phenomenal urban growth it would ever have. Hundreds of small cities and some larger cities emerged from frontier settlements or wilderness. Chicago's population grew from less than 100 in 1830 to over 110,000 in 1860; its downtown property value had exceeded $50 million by 1856 (at least ½ billion in 1983 dollars). At the same time, established cities such as Boston, Philadelphia, Baltimore, and New York became major economic and population centers. In 1810, the United States had no cities above 150,000 persons. In 1860, it had nine of them.

Because even the most technologically advanced cities in America were tightly packed "walking" cities as late as the 1850s, the population explosion of 1820 to 1860 threatened to become a sanitary disaster. From necessity, larger cities made their first major capital expenditures on water and sanitary facilities—facilities that did not directly promote city economic development (except their engineering and construction sectors) but were essential to its continuation (Warner, 1972: 202-203; Peterson, 1982).

Although the South before the Civil War was less than 10 percent urbanized versus the Northeast's 35 percent and the Midwest's 13 percent, southern cities and states had actively sponsored transportation improvements. And many southern urban economies grew impressively from 1830 to 1860. Still, northern merchants and factors had come to dominate much of southern trade. And Northeast-Midwest railroad and economic links by the 1850s were superseding South-Midwest waterway and economic links, greatly increasing sectional tension over the placement of federally subsidized railroads and intensifying southern fears about relative economic and political decline within the Union.

The "unconditional surrender" of the South after four years of civil war had as one important consequence the effective control of urban Northern finance and industry over southern economic life well into the twentieth century. Although southern cities shared in the national industrial growth of 1870 to 1929 and fared much better than the rural South, southern urban growth was probably restrained after the 1890s by northern control—by decisions, for instance, that slowed the steel-producing potential of Birmingham in favor of Northern steel centers. Only after 1945—in part because of substantial federal expenditures in the South that began in the 1930s and expanded in the war and postwar years—did southern cities emerge from northern dominance and contribute many of the growth centers of the 1970s—a historical turn of events that heightened the rivalry between cities and their regions after the mid-1970s (Brownell and Goldfield, 1977; Watkins and Perry, 1977).

THE GREAT CITY ERA, 1870 TO 1929

The period 1870 to 1929 is the major phase of the "great city era" of American history (Peterson, 1982). The agglomeration economies of centralized transport, markets, labor supply, communications, and applied technology were at their height. The great cities of 1870 extended old commercial advantages, and also developed services and industries with national markets. Smaller trading cities such as Detroit became great centers of new industries. And hundreds of smaller cities became influential midsized cities that dominated some industrial subsector. With their great railroad terminals and then with their skyscrapers, the cities symbolized American economic greatness. They also represented the main economic change for millions of migrants from rural America and Europe who provided manpower and additional demand for urban goods and property.

Because essential railroad and water connections were decided for all but a few Western cities by the 1870s, cities did not have to compete as aggressively as before 1870 for state and federal transportation investment. Even substantial federal funding of river and harbor improvements from the 1870s to 1920s did not affect much the relative standing of city economies.

Except for modernization of harbor and railroad facilities in some port cities, the major city investments of 1870 to 1929 were reserved for the core city and its territorial expansion. And these investments were financed by major borrowings of the cities or city-franchise from the more developed capital markets that followed the Civil War. In these financial dealings, the emerging political machines and bosses of larger cities preferred a minimum of state or federal interference in order to monopolize the political and personal advantages of major capital improvements. The most important demands of local leaders from state legislatures were political—greater rights of self-rule and borrowing and annexation rights over developed or developing areas on their fringe. The annexation rights were usually granted automatically until the early twentieth century because the city-supported utilities and transportation companies enabled much development and because the older city sometimes gave tax concessions and additional services to the annexed areas (Elazar, 1967; Olson, 1980: 217-221).

In addition to direct city borrowing, city leaders of the 1870s to 1900s financed urban infrastructure and its very high capital costs by giving monopolies, land grants, and tax concessions to transport and utility companies. "Honest" and "dishonest" graft may have shaped many of these inducements. But city leaders also saw great opportunities in linking major economic nodes of the core city and in opening surrounding areas to residential and commercial development. In fact, the revolution of transport, utilities, and communications within cities from 1870 onward greatly increased the property values of the developed areas (which were annexed) and also of the downtowns that became more specialized as commercial, service, and governmental centers. Even before the Chicago Exposition of 1893-1894 made city beautification and planning a Movement, some cities had also funded great parks and boulevards that boosted city self-esteem and multiplied property values in emerging areas of settlement.

In addition to developing land and raising the tax base, the great urban investments of 1870 to 1900 lessened some of the diseconomies of the agglomerated city. First, the transport systems and park projects lessened some of the unhealthy congestion of the core city, at least for the middle classes and skilled working classes. Second, major additional investments in sewerage and water systems warded off the worst dangers of urban congestion. Third, by using city finances as an informal public works and welfare system, the city bosses lessened some of the suffering and potential protest of the city laboring classes. Until the 1900s, however, urban bosses did not support reform of housing or working conditions.

The payrolls and graft of urban bosses were considered excessive by civic reformers of the 1890s and onward. These reformers wished to bring new professional and business principles of efficiency, and sometimes a dose of native Protestant uplift, to large government organization. But they also surpassed the commitment of the urban bosses to major capital improvements (McDonald, 1982: 361-362). Whether they won power or influenced local decisions, the civic reformers emphasized large-scale planning and the physical enhancement of urban sites in the capital improvement agenda. One of their causes, the City Beautiful Movement of the 1900s to 1920s, was intended to enhance the downtown by civic and cultural monuments, boulevards, and parks that replaced or controlled unsightly shanty areas and industrial uses. It was one harbinger of urban renewal.

At their zenith in the 1920s, with a massive growth of office towers in their downtowns, the major cities nevertheless had problems that only a booming national economy offset. They had accumulated high levels of debt to pay for public improvements. They had steadily lost the ability to annex the new residential suburbs, as middle-class residents wished greater self-control and now had the automobile as well as commuter railroads to enable living outside the central city. Central cities were also beginning to lose industrial jobs to suburbs that enjoyed cheaper and more efficient land arrangements. Some northern cities in the 1910s and 1920s also saw a shift of their textile and shoe industries to the South with its cheaper natural and human resources.

THE DEPRESSION DECADE, 1929 TO 1939

American cities in 1929 had no systematic relations of consequence with the federal government. Individually, they appealed less to the federal government than they had a century earlier. Still, Daniel Elazar (1967) has argued that cities in the 1920s had important informal links with the federal government and that the growing economic and geographic complexity of the major cities would have deepened their direct federal ties. But it was the Great Depression that brought cities as a group to the federal government and brought the federal government to the cities. Even after 1929, the federal government moved slowly in response to city economic distress and had important reservations when it moved in force.

The state governments to which cities first appealed were unwilling to support relief or work projects for residents of their cities. The cities themselves stepped up their public works and then greatly expanded direct relief. But drastic reductions in property tax collections and the sizable interest payments due on their bond obligations limited their capacities. In fact, overall city spending fell 15 percent per capita from 1929 to 1933, and much more in categories other than interest payments and relief. As part of their retrenchment, cities drastically cut their planning boards—one symbol of a common view during the early 1930s that the collapse of the private economy was a natural disaster beyond rational control.

Desperate for short-term relief, 26 mayors met in Detroit in May 1932 and appealed to the federal government for $5 billion to provide fiscal relief and to finance public construction projects. This group was the nucleus of the United States Conference of Mayors, which formally organized in 1933 and became an effective urban lobby during the Roosevelt years. Herbert Hoover, however, rejected unconditional fiscal aid and rejected public works projects that lacked a private economic rationale. Hoover's major response to the Depression, the Reconstruction Finance Corporation, primarily shored up the shaky finances of large private companies, especially banks and railroads. It failed to stabilize investment or job levels. RFC loans to local governments were restricted to self-liquidating projects and did not significantly help city economies or finances. Even under New Deal direction, the RFC was ultra-cautious in making loans to cities.

In his First Hundred Days, Roosevelt tried to limit the destructive deflation and competition of banking, industry, and agriculture. More creatively, he proposed major conservation efforts and a Tennessee Valley Authority to develop wasted natural and human resources. All of these proposals lacked an urban focus. Indeed, the specific development programs of the New Deal years (TVA, rural electrification) were primarily southern as well as rural—as would be the regional economic development programs of the 1960s (Levitan, 1976). Of course, the rural South had long been the most economically backward area of the nation. And a major public investment program in the South would create industrial orders for the urban North as well as raise the productivity of southern human and natural resources. But also motivating Roosevelt in his development choices was an emotional preference for the countryside. Despite his many years of association with city politicians, Roosevelt, according to Rexford Tugwell of his brain trust, "always did and always would regard the cities as rather hopeless" (Miller, 1979: 13). In fact, the most specific urban proposals that Roosevelt authorized would have involved migration from established cities to the countryside (Gelfand, 1975: 25-26). Only the realities of high capital costs and the continued rural migration to the cities curbed these plans.

Even if Roosevelt "would not commit himself or the Federal government to helping the cities themselves", he was "committed to helping people who lived in the cities" (Miller, 1979: 13). He also came to see city economic stabilization as important to national economic stabilization, and he most definitely came to appreciate city votes. Therefore, he authorized a series of relief and public works programs that increasingly helped cities. To appreciate their magnitude, 1930s values given below should be multiplied by seven to approximate 1983 values. The 1933 Federal Emergency Relief Act provided $.5 billion in relief funds. Against city objections, states were given the funding decisions and alloted a disproportionate amount to rural areas. But under the Civil Works Administration (1933-1934, $1 billion), the Public Works Administration (1933-1939, $6 billion, 45 percent in grants and 55 percent in loans), and the Works Progress Administration (1935-1943, $11 billion, 25 percent local contribution after 1939), the federal administrators who allocated funds consulted city interests and gave cities the largest share of funds. The cities, in fact, lobbied to keep the allocations from Congress.

Although cities in all regions received major benefits from these programs, older and politically influential cities fared well. New York City was especially well funded because of the political energies of Mayor Laguardia and the planning energies of Robert Moses. In addition to their primary task of taking the unemployed off relief rolls, these programs offset some of the cities' drastic cuts in capital expenditures and funded airports, sewers, and roads as well as schools and parks. Despite city lobbying to maintain PWA investments, Roosevelt gave in to conservative pressure and let the program lapse before city economies were fully restored.

The New Deal agency that most affected the housing industry and long-term urban development patterns was the Federal Housing Administration of 1934, which insured long-term loans for families to renovate homes and especially to purchase new homes. During the 1930s, FHA programs undoubtedly aided some city properties and many city-based construction firms. But after 1945, the tilt of FHA mortgages toward less risky new construction in suburban areas led to residential disinvestment and to a weakened tax base in the central city. Miller (1979: 13) judges that the "role of the FHA in the residential abandonment of central cities was recognized from the start." But it accorded with the general New Deal philosophy that the city resident, not the city, should be the focus of federal concern (Gelfand, 1975: 68). Residents left behind were not considered.

From 1937 to 1941, during the waning period of the New Deal, two groups working in the Department of Interior focused upon city problems and proposed a broad federal commitment to improving city fiscal, economic, and resident condition. Neither group was influential, even though an uncle of Franklin Roosevelt led one group. Their major works—*Our Cities, Their Role in the National Economy* (Committee on Urbanism, National Resources Committee, 1937) and

Action for Cities: A Guide for Community Planning (National Resources Planning Board, 1942)—might be seen as embryonic urban policies that never developed.

THE ERA OF URBAN RENEWAL, 1940 TO 1965

New Deal housing programs of slum clearance and public housing treated slums in the traditional way as threats to the health, safety, and morals of their residents. But as the great downtown improvements of the 1920s gave way to deferred maintenance in the 1930s, and as the overall city tax base became more precarious, urban real estate and business groups saw slums as a "blight" on city property values and finances. In 1941, the Federal Housing Administration followed the lead of an influential lobbyist for these views, the National Association of Real Estate Boards (NAREB), and studied the "possible redevelopment on an economic basis of substandard blighted areas." Its resulting *Handbook on Urban Redevelopment of Cities in the United States* emphasized local methods of altering property uses, but conceded that federal assistance might be needed for city reconstruction.

Before the federal government legislated urban redevelopment in 1949, cities and states had begun to act. In 1941, the New York State Legislature heeded New York City merchants and passed an Urban Redevelopment Act that "allowed localities to grant tax exemptions and the power of eminent domain to private corporations engaged in slum clearance and rebuilding." Developers, however, were limited to a 5 percent rate of profit. Another thirteen states passed some form of urban redevelopment legislation by 1945, and twenty-five altogether by 1948. These acts spurred little immediate activity, however, because of profit caps and because of no supporting subsidy (Gelfand, 1975: 136-137). An exception was Pennsylvania's 1946-1947 redevelopment package tailored for Pittsburgh; it gave the city liberal powers of eminent domain and specifically permitted insurance companies to invest in cleared areas. Pittsburgh earned these broad powers and by the mid-1950s had quickly achieved phase one of the first of the so-called urban renaissances because of its rare combination of qualities—a tradition of planning by civic and private groups to stem the downtown's obvious stagnation and gritty image, the local pride and investment capital of its major corporations (galvanized by the wealthy and influential Richard King Mellon), tracts of underused commercial and industrial land next to the compact downtown, and the judicious leadership of Mayor David Lawrence (who guided legislation and a $1 million city land write-down, but left actual planning to the private sector). (Pittsburgh's renewal, past and present, is summarized well by the case study in Fosler and Berger [1982].)

During the war years and the immediate postwar years, many older cities enjoyed high employment and property values. On the other hand, their downtowns and housing stock had been somewhat neglected for almost two decades. After 1945, the cities were clearly losing middle-class families to the suburbs. And though the feared postwar depression never arrived, the shadow of the 1930s lingered. The cities could not take for granted a booming national economy that lifted all areas, nor could Congress and the executive branch dispense with new urban investment in supporting national investment and job levels. Thus, after several years of preparation, the National Housing Act of 1949 made important concessions to city real estate and business interests that emphasized the clearance of blighted areas for alternative land use rather than for the rehousing of slum residents. Except for some rural congressmen who objected to using taxes to save cities from their own mistakes, Congress did not actively debate the Title I Redevelopment part of the Housing Act (Gelfand, 1975: 145-156).

Title I of the Housing Act gave primary responsibility to local agencies in determining the alternative uses of their redeveloped areas, encouraged the role of "free enterprise" in such redevelopment, did not cap private profits, and did not require that the low-cost public housing authorized in Title II of the Act be placed in redeveloped areas. In short, the federal government sanctioned local authorities to use eminent domain to obtain land not only for public housing (as in the 1937 Housing Act), but also for profitable development by private developers. Equally important, the 1949 Act authorized $1 billion in loans for redevelopment planning and land acquisition and $.5 billion in long-term capital grants to pay two-thirds the difference between the cost of preparing the land for reuse and the actual resale price of the land to the developer. Reuse costs included public works improvements as well as land purchase and displacement fees (the latter were slight until the last years of the program in the late 1960s and early 1970s).

Despite its apparent generosity of powers and funds, Title I was not used extensively by cities until the mid-1950s. One reason is that Title I had restricted grants to blighted areas that had been primarily residential or were to become primarily residential. (For the background of this provision, see Wilson, 1967: 104-107.) Even when they were not skeptical about government programs, large-scale developers were reluctant to invest in areas that were recently cleared of slums and blight. Moreover, local and federal officials were still learning how to implement the Act flexibly.

Despite its shaky start, Title I activity expanded quite steadily after the mid-1950s. On the demand side, downtown merchants and property owners now felt the full adverse impact of suburban migration and malls and clamored for downtown revitalization. City mayors who had to contend with declining city property values and the increased costs of servicing low-income migrants saw transformed land as a fiscal panacea. Nonprofit institutions such as universities and medical schools felt constricted, if not threatened, by enclosing urban slums; they saw renewed land as their best chance to modernize facilities and attract clients. Corporate giants, especially banks and insurance companies, wished to preserve their past urban investment and saw opportunities for new urban investment that could supply essential headquarters space and monuments to their glory (rather like Florentine towers of the Renaissance). They too began to appreciate the virtues of developing marked-down land near choice locations.

Matching the greater interest in redevelopment funding was the greater experience of federal and local officials in implementing the Act, increased funding levels (especially from 1961 to 1967), and a redirection of the Act toward economic and fiscal development. Miles Colean, a housing economist who influenced the 1954 Urban Renewal Act that amended the 1949 Urban Redevelopment program, aptly expressed the program's redirection:

> The renewal problem is primarily one of how to construct, maintain, and rebuild the various parts of the urban structure so that the city as whole remains at all times in a sound economic condition from the point of view of both the private and public interest [Gelfand, 1975: 172].

An immediate consequence of the more explicit economic approach was the exemption of 10 percent of the grants in the 1954 Renewal Act from the primarily residential requirement—an exemption raised to 20 percent in 1959, 30 percent in 1961, and 35 percent in 1967.

Long before its official termination in 1973, the Title I Urban Renewal program had become the most extensive and controversial urban development program ever undertaken by the federal or local governments. In congressional hearings of 1979 (Committee on Banking) on lessons of past urban economic development programs, the Urban Renewal program of the late 1950s and 1960s was pilloried as a wasteful anachronism or praised as a liberator of urban economic energies. Before generalizations are made, the context and controversial legacy of the urban renewal experience will be conveyed by a summary of Boston before renewal and of three of Boston's most ambitious projects in the heyday of renewal.

Boston in the mid-1950s seemed a faded dowager. Its labor-intensive industries had long been undercut by cheaper areas in the South and abroad. Its antiquated land and street patterns could not compete with the suburban space near radial highways such as Route 128 (focus of a recent electronics research boom). Its older wharves, railroad yards, and public market areas testified only to past economic greatness and public-private innovativeness. Its extremely tight boundaries and older housing stock were invitations to a large suburban outmigration. Even its office sector seemed outmoded and cramped; Boston had the same square footage in 1945 as in 1925 and only about one million feet more by the late 1950s—an opportunity, perhaps, but a sign of stagnation. Boston's property tax burden was among the nation's highest and rising. Boston's financial and political leadership saw a sagging tax base as Boston's key problem and urban renewal as the solution (Wilson, 1967: 257-263; Conzen and Lewis, 1976: 50).

Planned from the early 1950s and carried out from 1958 onward, the West End residential renewal project became an infamous symbol of the bulldozer excesses of Urban Renewal in its first flush. The Boston Renewal Authority was determined to raise residential property values and to bring back upper-income commuters from the suburbs. With the promise of extensive federal write-down subsidies for later land transfer, the Boston authority used its formidable condemnation powers (with the approval of the Mayor and City Council) to raze a densely populated, viable ethnic neighborhood with an excellent location near the river, downtown, and expressways. As a result of threatened and actual bulldozing, about ten thousand residents and numerous neighborhood businesses were displaced with little or no compensation for moving expenses, higher rents, and loss of customers. The loss of personal and community ties was intense and was also overlooked by redevelopment officials. What counted for these officials was that within a few years, high-rent apartment buildings would be built that raised property values considerably and provided workers and customers for downtown office and retail projects (Gans, 1962: 283-328; Conzen and Lewis, 1976: 69-71).

Boston's nonresidential renewal in the late 1950s and early 1960s involved much less displacement and much clearer net economic benefit. For its first major commercial project, the Prudential Center, the Boston Authority enabled land assembly of old railroad yards and their replacement by a massive office and apartment complex that initiated Boston's skyscraper image. Private capital financed everything. The only public "cost" was a city tax abatement for forty years. Yet, in a decade, the city increased its taxes for the area from an estimated $.75 million to an estimated $3 million per year because of $150 million of private investment. The center also brought 7,500 jobs to the area (4,100 new to the city), of which an estimated 62 percent were held by commuters (Perloff, 1975: 160-161). Boston's next major office project, the Government Center, used federal and city funds to clear a seedy commercial area next to the downtown and to replace it with a strikingly modern city hall, an office complex, parking spaces, and open plazas. State and federal buildings soon also embellished the area. Spurred by renewal projects such as these, Boston's office space almost doubled from 1957 to 1973, and its employment in office sectors rose over 50 percent. This economic vitality, in turn, facilitated in the 1970s the creative reuse of old markets and wharves and a construction boom of offices, hotels, and highrise condominiums in or near the downtown.

Certain aspects of the Boston renewal experience can be generalized. Urban Renewal areas in or near prime downtown land had an impressive growth of commercial and institutional investment. This investment led directly to much higher property values and taxes, even after discounting the loss of taxes while land was cleared (Lineberry and Sharkansky, 1978: 378-379; Bingham, 1974). The investment also led directly to a substantial growth of jobs, though primarily for skilled professionals who often resided outside the city. The net effect of urban renewal on city economic expansion can be argued. After all, these urban renewal successes accorded with a national growth of white-collar, service, medical, and educational sectors—sectors that had long favored central city locations. Still, their continued loyalty and expansion in older central cities was not a foregone conclusion in the early and mid-1950s. A broad range of city development specialists have testified to the importance to commercial and institutional revitalization of urban renewal's critical mass of powers, subsidies, and public attention (Committee on Banking, 1979; Fosler and Berger, 1982; Bellush and Hausknecht, 1967: 400-406).

Clearing lower-income residences for middle- and upper-income residences involved the greatest amount of land, deep subsidy, and controversy. Sites with a good location near the downtown or institutions were developed quite rapidly and yielded net tax advantages within a decade. Other sites were undeveloped for years, although some development officials felt that a removal of their worst housing and its high densities of troubled households was a sufficient reason. Still, the removal of thousands of low-income rental units from fairly tight housing markets imposed large and uncompensated costs on residents (Gans, 1962). Massive clearance of neighborhoods might also have destabilized other neighborhoods and caused some of the costly urban disorder of the 1960s (Committee on Banking, 1979; Rothenberg, 1967: 223-229). Only after 1967 were federal guidelines able to make local Urban Renewal programs reasonably sensitive to lower-income housing needs (Sanders, 1980).

Retail development under urban renewal was not nearly as extensive as office, institutional, or residential development. The needs of downtown merchants were usually met indirectly by revitalizing nearby commercial and residential areas and by providing more parking space. The lure of suburban retail sites was so great that investors saw too much risk in new retail developments. The most successful retail renewal of the 1950s, the Nicollet Mall in Minneapolis, stressed rehabilitation, walkways, street improvements, and nearby parking in a cleared area rather than new retail stores (Abler, 1976: 44-46). But urban renewal land cleared for large-scale retail projects often stayed vacant, unintentionally becoming a useful "landbank" for UDAG commercial and retail projects many years later (Committee on Banking, 1979; Nathan and Webman, 1981).

With some notable exceptions such as Philadelphia's Eastwick project, large urban renewal projects in older cities were not industrial. The civic and business leaders of these cities did not show a special concern with the industrial job base. It did not fit their image of a modern city able to compete with the suburbs for buildings, jobs, and residents. In addition, the primarily residential requirements of urban renewal and the difficulty of finding large tracts of land suitable for industrial purposes further inhibited a use of funds for industrial projects. Finally, industrial projects usually lacked the extra financing advantages of mortgage-guaranteed projects.

In addition to dramatizing the priorities of urban renewal in larger cities, the Boston renewal experience highlights an innovation in public service. Public officials such as Logue of Boston and New Haven, Mayor Lee of New Haven, Danzig of Newark, and Moses of New York were "public entrepeneurs" with skills in matching land and public subsidies to developers, with a zeal for large and simultaneous projects that reshaped urban economic land use, and with a concern for the civic bottom line that overrode the needs of some of their "less productive" constituents. Not many renewal officials had their influence, but the urban renewal experience undoubtedly created a group of public officials and private developers who were alert to the role of both the federal and local government in shaping local economies (Bellush and Haussknecht, 1967: 209-225, 239-255).

Boston also symbolizes well the disproportional funding of urban renewal. Renewal funding was categorical and large, so that the most active renewal directors and mayors could easily persuade federal officials of the value of their projects. In the first decade, when federal officials were desperate to get action, New York City under Robert Moses garnered a remarkable share of the construction grants. The very high per capita funding of Boston, New York, and New Haven also illustrates the concentration of urban renewal funds on large and older central cities in the North. These cities had clear opportunities for improved land use and the motivation to empower dynamic officials who could amass federal funding. In contrast, smaller cities usually lacked the need or the planning resources. And many of the large but newer cities of the Sunbelt had downtowns more adapted to the automobile age, were able to annex the tax base of their fringe areas, and had public and private leaders who were skeptical of federal interference in local economic affairs (defense and space contracts were another matter). Despite exceptions to these tendencies—after all, one thousand cities had participated in Federal Urban Renewal by 1973—their aggregate effect was that more than 55 percent of the $10 billion in Federal Urban Renewal grants went to cities with more than 100,000 persons and that cities in Massachusetts had almost twice the grants of the combined cities of Arizona, Florida, Louisiana, New Mexico, South Carolina, and Texas (U.S. Department of Housing and Urban Development, 1974).

THE RETREAT FROM ECONOMIC DEVELOPMENT, 1966 TO 1974

Urban renewal's heyday as a development program was the mid-1950s to the mid-1960s. Although the Federal Renewal program made new grants until 1973, the main attention of the Federal government and of many localities from the mid-1960s to the mid-1970s was direct assistance to individuals, to neighborhoods as social units, and to general city revenues. This shift of emphasis resulted from the social and economic problems of massive urban renewal, from the ghetto riots of the 1960s, and from the special governmental emphases of Presidents Johnson and Nixon (Committee on Banking, 1979).

The more experimental programs of training disadvantaged persons and improving the social as well as physical status of neighborhoods lost much of their governmental backing in the early 1970s. But the great increase of monetary and in-kind entitlements to poor persons and the introduction of broad purpose entitlements by the federal government to local governments became benchmarks of the 1970s. And these programs went against the grain of classic urban renewal, which emphasized large-scale physical redevelopment in a fairly restricted number of places. Already much diffused by the early 1970s, Urban Renewal was just one of many activities folded into Community Development Block Grants in the Housing and Community Development Act of 1974.

MAKING THE MOST OF SCARCITY, 1975 TO 1980

The contrasts of 1975 to 1980 with earlier periods of urban economic practice are sharp enough that the period is historically more defined than a passing current event. The period started with the severest postwar recession up to that time, a recession that hit Northern central cities with special force and coincided with a full appreciation of their sharp job losses after 1969; the near-bankruptcy of New York City and fiscal stringency in other major Northern cities; a decline in population of some metropolitan areas as well as cities; mounting evidence of a crumbling infrastructure; and massive residential abandonment that has been sardonically called urban clearance of the private sector without any renewal. The period concluded with sustained high rates of inflation and interest, a boom in prime real estate, a moderate recovery and sharp downturn, local taxpayer revolts, and federal restraint in domestic programs.

The distinct set of historical forces contributed to new patterns of public investment and urban economic development in this period—a forceful and self-conscious competition for federal development resources; much greater concern for macropolicy as the context of local development; greater selectivity and resourcefulness in planning development; greater attention to the city past as a handicap and opportunity; and a return to the pre-1965 emphasis on physical investment in places rather than investment or subsidies for persons.

Just as job losses and fiscal strain were prompting them to seek additional development resources from Washington, declining cities faced stiffer competition from faster growing cities that had become accustomed by the Block Grant revolution to federal assistance and wished to maintain their advantageously low tax rates. The competition of cities was joined to and stiffened by the competition of regions; the recession of 1974-1975 had battered some regions and left others almost untouched; the regional convergence of industrial shares and of per capita income that had been perceived favorably or neutrally from the 1920s to the late 1960s was now feared to be a regional divergence, with the once-dominant North and Midwest and especially their cities allegedly sliding into economic decline (Advisory Commission on Intergovernmental Relations, 1980). The regional rivalry had an urban-rural twist when older central cities centered in the North claimed that they needed economic development funding more than fast-growing rural areas centered in the South. The rural representatives replied that their economic growth had lifted them only partly from their historic economic backwardness.

In terms of costs to the treasury and total activity (though not net activity), an escalating and varied use of industrial development bonds by all types of cities was the major effect of the new competition. The most publicized effect was the addition to Community Development Block Grants of a formula that assured older cities a share of funding nearer to their categorical program shares of the late 1960s and early 1970s. (Because of a "hold harmless" clause, older cities had kept their categorical funding levels up to 1977.) Urban areas also won a greater share of EDA funding. Both redistributions were facilitated by larger allocations for both programs in the first Carter years. After 1978, development funding and redistribution were both curtailed. It is even arguable that the bitter competition between areas for preference in the self-conscious era of computer printouts might have weakened the political coalition for new sources of funding such as a National Development Bank. (Inter-agency rivalry, budget constraints, and doubts about the bank's effectiveness also doomed this large-scale development initiative.)

Limits on new development programs, the growth of regional research groups, the introduction of urban impact statements, and the research needs of academic urban economists all increased the volume and specificity of research on the broad effects of federal spending and taxing policies on urban areas and regions. Researchers (Vernez, 1979; Advisory Commission, 1980; Chinitz, 1980: 67-78; Glickman, 1980: 19-43, 81-88, 451-54) usually concluded that tax policy, indirect development programs such as highway and sewer construction, defense payrolls and contracting, and probably energy pricing and environmental regulation had much more impact on urban economic growth than specific economic development programs. On the whole, these impacts were said to work against older central cities and also against older regions, although the distribution of defense contracts favored some New England states and could also be justified by the cost-competitiveness of the chosen areas (Mieszkowski, 1979). The Carter administration or Congress softened some of the unfavorable impacts with changes in procurement policy and in guidelines for mortgage insurance and wastewater facilities, a shift of highway funds to mass transit, reforms of tax legislation to encourage rehabilitation and historic preservation, and a stretching out of the decontrol of natural gas prices. The changes and their impact were modest (Chinitz, 1979: 67-78; Webman, 1981; U.S. Department of Housing, 1980).

The "targeting" of scarce resources to the neediest jurisdictions was the most publicized approach to selectivity, but not the most widespread or influential. At the local level, selectivity usually connoted a search for probable successes and the maximum job or fiscal payoff—types of selectivity that often overlooked the neediest neighborhoods or persons. The decrease of long-term federal money for planning and investment and the aftershock of Urban Renewal led to the type of pragmatism that one veteran of renewal unhappily summarized: "Make no big plans because we can't alter the key forces built into the existing urban development process, and we had better concentrate on small-scale projects and small fundings on things that can be done with short term horizons, where we have a better probability of success" (Committee on Banking, 1979). Economic development handbooks suggested, more cheerfully, that cities should research and build upon their remaining advantages, carefully inventory and guide the economic uses of their all-important land parcels, and listen carefully to the complaints of their businessmen.

Even if the 1975 to 1980 period rarely had public development projects or directors to match those of the urban renewal heyday, it had its own elements of creativity. In an era of multiple federal and state programs, of high interest rates, of inflated real estate values, and of complicated tax write-offs, city development staff learned to assemble a variety of grant, loan, and tax inducements as well as available land. Moreover, for choice sites, development officials became more adept at estimating and scheduling the payoff to the city. And for sites with a large city investment such as convention centers or mass transit nodes, cities developed competitive bidding in which a broad range of direct and spillover benefits were judged. Cities wished to avoid high-value but sterile office towers confined to working hour uses. Indeed, city development officials became so absorbed in crafting their downtown projects that some neighborhood groups of the late 1970s accused them of becoming economic dealers attentive only to the prime site (Fosler and Berger, 1982).

With antecedents of demolition and occasional renovation under urban renewal, the older urban landscape burst forth as an obstacle and as a resource from the mid-1970s onward. The age of a city became an intellectual shorthand for its institutional flexibility, boundary and fiscal constraints, job skills, technological level, and state of its infrastructure (Watkins, 1980; Norton, 1979; U.S. Department of Housing and Urban Development, 1979b: 105-116; Chinitz, 1979: 15-45). Older central cities were viewed, sometimes deterministically, as troubled or distressed—even cities such as Minneapolis and San Francisco with a high growth of per capita income and of per capita property values were so classified. Instead of relying on age of housing as a measure of city economic and fiscal need, a classification of need used in some Housing and Urban Development analyses has emphasized rates of poverty and unemployment and the net and percentage growth of per capita income (Benjamin et al., 1983; U.S. Department of Housing, 1980). These measures were first used to moderate the effects of age and decline indicators in the Urban Development Action Grants (U.S. Department of Housing, 1979a).

Public policy sought ways to help cities "grow old gracefully" (Committee on Banking, 1977). Most directly, the Community Development Block Grant Formula revision of 1977 gave a significant funding role to age of housing and to an associated variable, population decline, that had the more dramatic effects on funding (U.S. Department of Housing, 1979b). Tax advantages for historic preservation and rehabilitation were increased, though new structures still benefited the most. Because of the tax changes, because of the difficulty of clearing older settled areas, and because of a public nostalgia for the much-diminished past, developers and city officials found investment resources in some of the older residential and commercial structures. These well-built, often elegant structures offset the gleaming sameness of central city highrises and suburban malls. In addition, downtown port areas that had once sustained city economies but had then turned derelict were now reclaimed for city and suburban shoppers alike. (Some of the most successful of these renovations such as Baltimore's Harborplace had Urban Renewal origins, however.)

Despite the much greater commercial appeal of old structures, the economic restoration of the past had its limits. Old structures that blocked efficient land assembly for high-priority in-town malls and great office towers were still viewed as obstacles, especially if their historic value was questionable. And old warehouse and factory lofts that were converted to residential and retail charm had the possible defect of foreclosing or hampering industrial uses that could employ less-skilled city residents.

As contrasted to the 1965 to 1974 period, the 1975 to 1980 period had a greater emphasis on the overall economic productivity of the city than on its neediest residents or on their neighborhoods. The shift of emphasis had several rationales: the great increase in federal and local assistance prior to 1974 to care for individual needs; the difficulty in training the very poor; the fiscal hardship of many cities, and the seeming lesson of New York that a city could not redistribute income. Moreover, the threat of ghetto riots had receded by the mid-1970s. Economic or fiscal distress cloaked the most dramatic alterations. As part of its restructuring, New York City cut the real value of its welfare and health expenditures and directed a much higher proportion of its budget to economic infrastructure. Less dramatically, CDBG amendments during this period also gave greater attention to city economic development, although well-organized neighborhood groups used other CDBG provisions to restrain the actual increase in economic development funding from 5 percent of entitled Community Development Block Grant funding in 1978 to 8 percent in 1982.

To add an economic development focus to the Community Development Block Grant program, the Carter administration and Congress initiated in 1977 the categorical Urban Development Action Grant program (UDAG). This much-publicized program reflects many of the recent trends of public investment in urban economies and will be briefly summarized. (UDAG will be covered more fully in a later chapter of this book.)

UDAG's statutory emphasis was on a city's overall economic, physical, and fiscal distress. Consequently, the selection procedures in the statute and in practice emphasize overall city need and leveraging of aggregate private investment, jobs, and tax benefits. Benefits to unskilled or unemployed workers are much less influential (Glickman, 1980: 335-362).

As a practical "action" program, UDAG has been noteworthy for enacting a strict prior commitment from the private sector and for working with local governments and developers on sophisticated leveraging and payback of public funds. UDAG was intended to avoid the fate of Urban Renewal and EDA sites that were cleared and prepared by government funding and then developed many years later or not at all by private developers. But critics of UDAG (found in Committee on Banking, 1979; Nathan and Webman, 1981) claim that UDAG projects involve little private risk, follow broad private trends, and often capitalize on sites prepared under the larger powers and resources of Urban Renewal.

Whatever its ultimate effectiveness in leveraging investment, jobs, and development skills for distressed areas, UDAG's emphasis on leveraging symbolizes well the public economy of scarcity in recent years. Its initial funding was $400 million—a "puny" amount to one skeptic—and its peak funding was $670 million, or about one-fourth the annual amount in constant dollars of Urban Renewal in its peak years. Even so, UDAG's distribution of funds has been contested from the start—by older, northern cities that gained a primary reliance of eligibility and need-based selection criteria on age of housing and population decline indicators; by smaller cities that won a 25 percent setaside of total funds; and by ineligible cities concentrated in the Sunbelt that in 1979 won Pockets of Poverty consideration for their needy subareas (U.S. Department of Housing, 1979a). UDAG also bans the use of grants to pirate jobs.

UDAG did not deal explicitly with macropolicy. But, implicitly, it was viewed as partial compensation for the harm of other federal actions on older central cities (Glickman, 1980: 335-362; U.S. Department of Housing, 1980).

A PARTIAL PROSPECTIVE

The "waterways to waterfronts" part of this chapter's title suggests that public investment in cities has shifted from physical essentials to "amenities," from the great transportation projects of the nineteenth century to the cultural, shopping, tourist, and convention inducements of the late twentieth century. Such a broad sweep omits a good part of the truth—public modernization or extension of facilities such as ports, mass transit, railroads, bridges, or airports still undergirds urban economic vitality. But the phrase does convey the economic importance of adapting to the much greater role of physical and social amenity in the purchasing and locational preferences of American consumers and suppliers. Whether all cities can and should adapt to these preferences, at the same time affording essential physical and human investments, is one of the hardest questions of public policy.

An alternative phrase, "from waterways to warships," would convey an equally important trend of public investment: the large-scale investments of defense. The respective roles of politics and economic factors in distributing defense contracts can be debated (Advisory Council, 1980; Mieszkowski, 1979). The overall efficiency of Defense spending on an entire economy can also be debated, as can the actual contribution and stability of past types of Defense and Space spending. But the large and growing proportion of government investment on Defense in the 1980s and its sizable effects on aggregate demand and technological spinoffs in urban areas should make it the economic development prize of the 1980s. Long known for its resourcefulness in maximizing its limited natural advantages, New England and many of its urban areas have enjoyed an economic turnaround during the past decade. And though excellent research facilities, physical amenity, creative reuse of older facilities, and greater attention to business needs have all worked together, good fortune in the great Defense race has also paid off.

REFERENCES

ABLER, R., J. S. ADAMS, and J. BORCHERT (1976) The Twin Cities of St. Paul and Minneapolis. Cambridge, MA: Ballinger.

Advisory Commission on Intergovernmental Relations (1980) Regional Growth: Historic Perspective. Washington, DC: Government Printing Office.

BELLUSH, J. and M. HAUSKNECHT [eds.] (1967) Urban Renewal: People, Politics, and Planning. Garden City, NJ: Doubleday.

BENJAMIN, R., H. BUNCE, and S. NEAL (1983) "Recent employment trends in large urban areas," pp. 277-305 in U.S. Department of Housing and Urban Development, Effects of the 1980 Census on Community Development Funding. Washington, DC: Government Printing Office.

BINGHAM, R. (1975) "Impact of federal grants on city government: public housing and urban renewal," 183-201 in R. Lineberry and L. Masotti (eds.) Urban Problems and Public Policy. Lexington, MA: D. C. Heath.

BROWNELL, B. and D. R. GOLDFIELD [eds.] (1977) The City in Southern History: The Growth of Urban Civilization in the South. Port Washington, NY: Kennikat Press.

CHINITZ, B. [ed.] (1979) Central City Economic Development. Cambridge, MA: Abt Books.

CHUDACOFF, H. (1981) The Evolution of American Urban Society. Englewood Cliffs, NJ: Prentice-Hall.

Committee on Banking, Finance, and Urban Affairs, Subcommittee on the City (1979) Urban Economic Development: Past Lessons and Future Requirements. Washington, DC: Government Printing Office.

———(1977) How Cities Can Grow Old Gracefully. Washington, DC: Government Printing Office.

CONZEN, M. and G. K. LEWIS (1976) Boston: A Geographical Portrait. Cambridge, MA: Ballinger.

ELAZAR, D. (1967) "Urban problems and the federal government: a historical inquiry." Political Science Quarterly 82 (December): 505-525.

FOSLER, R. S. and R. A. BERGER [eds.] (1982) Public-Private Partnership in American Cities: Seven Case Studies. Lexington, MA: D. C. Heath.

GANS, H. (1962) The West End: An Urban Village. New York: Macmillan.

GELFAND, M. I. (1975) A Nation of Cities: The Federal Government and Urban America, 1933-65. New York: Oxford University Press.

GLICKMAN, N. J. [ed.] (1980) The Urban Impacts of Federal Policies. Baltimore: Johns Hopkins University Press.

GOODRICH, C. (1960) Government Promotion of Canals and Railroads, 1800-1890. New York: Columbia University Press.

LAMPARD, E. E. (1968) "The evolving system of cities in the United States: urbanization and economic development," in H. Perloff and L. Wingo, Jr. (eds.) Issues in Urban Economics. Baltimore: Johns Hopkins University Press.

LEVITAN, S. (1976) Too Little But Not Too Late: Federal Aid to Lagging Areas. Lexington, MA: D. C. Heath.

LINEBERRY, R. and I. Sharkansky (1978) Urban Politics and Public Policy. New York: Harper & Row.

McDONALD, T. (1982) "From economics to political economy in the history of urban public policy." Journal of Urban History 8 (May): 355-63.

MIESZKOWSKI, P. (1979) "Recent trends in urban and regional development," pp. 3-39 in P. Mieszkowski and M. Straszheim [eds.] Current Issues in Urban Economics. Baltimore: Johns Hopkins University Press.

MILLER, R. M. (1979) "The federal role in cities: the New Deal years." Commentary (CUED) 3 (July): 11-13.

NATHAN R. and J. WEBMAN [eds.] (1981) The Urban Development Action Grant Program. Princeton: Princeton Urban and Regional Research Center.

NORTON, R. D. (1979) City Life-Cycles and American Urban Policy. New York: Academic Press.

OLSON, S. (1980) Baltimore: The Building of an American City. Baltimore. Johns Hopkins University Press.

PERLOFF, H., T. BERG, R. FOUNAIN, D. VETTER, and J. WELD (1975) Modernizing the Central City. Cambridge, MA: Ballinger.

PETERSON, J. (1982) "Environment and technology in the great city era of American history." Journal of Urban History 8 (May): 343-354.

ROTHENBERG, J. (1967) Economic Evaluation of Urban Renewal. Washington, DC: Brookings Institution.

SANDERS, H. (1980) "Urban renewal and the revitalized city: a reconsideration of recent history," pp. 103-126 in D. Rosenthal (ed.) Urban Revitalization. Beverly Hills, CA: Sage.

U.S Department of Housing and Urban Development (1980) The President's National Urban Policy Report. Washington, DC: Government Printing Office.

———(1979a) Pockets of Poverty: An Examination of Needs and Options. Washington, DC: Government Printing Office.

———(1979b) City Need and Community Development Funding. Washington, DC: Government Printing Office.

———(1974) HUD Statistical Yearbook, 1974. Washington, DC: Government Printing Office.

VERNEZ, G., R.J. VAUGHN, and R. K. YIN (1979) Federal Activities in Urban Economic Development. Washington, DC: Economic Development Administration, U.S. Department of Commerce.

VERNON, R. (1959) The Changing Economic Function of the Central City. New York: Area Development Committee of CED.

WARNER, S. B., Jr. (1972) The Urban Wilderness: A History of the American City. New York: Harper & Row.

WATKINS, A.J. (1980) The Practise of Urban Economics. Beverly Hills, CA: Sage.

———and D. PERRY (1977) "Regional change and the impact of uneven urban development," in D. Perry and A. Watkins (eds.) The Rise of Sunbelt Cities. Beverly Hills, CA: Sage.

WEBMAN, J. A. (1981) "The Carter administration and urban policy." Presented at the American Political Science Association meetings, August.

WILSON, J. Q. [ed.] (1967) Urban Renewal: The Record and the Controversy. Cambridge, MA: MIT Press.

2

Urban Growth and Technological Innovation

ROBERT PREMUS

☐ URBAN ECONOMISTS have long recognized the importance of cities as contributors to the nation's technological development (Hoover, 1975: 238-244). Cities have historically served as incubator centers from which many of the nation's major technological innovations emerged. These technological innovations contributed significantly to national economic growth and prosperity. Cities grew and developed in the process as people and jobs shifted from rural to urban regions. Yet, in spite of the widespread recognition of the importance of cities as technology centers, few communities looked to technological innovation as a source of regional economic growth. The notable exceptions are Route 128 in Boston, the Silicon Valley in San Francisco, and Research Triangle Park in the Raleigh-Durham-Chapel Hill area. It is fair to say that most urban communities preferred to conceptualize their communities as manufacturing and service centers, and economic development as an expansion of these sectors.

An important question confronting urban communities today is whether or not they can afford to continue to neglect the important role technological innovation can play in the development of their regions. Most urban communities can no longer expect to significantly benefit from technologically induced urbanization. For these urban areas, if not for all, future growth and prosperity will depend much more on their ability to innovate and incubate new industries and less on an extension in their existing economic base.

A growing awareness of the new competitive realities confronting urban centers is probably behind the rapid pace at which urban

AUTHOR'S NOTE: *The author is an economist with the Joint Economic Committee, Congress of the United States, on academic leave of absence from Wright State University. Alexis Stungevicius is to be thanked for her assistance on the chapter, although the author assumes full responsibility for any errors that may have been overlooked.*

communities are developing high-tech strategies. The purpose of this chapter is to examine how these cities are responding to the opportunities and challenges presented to them by technological change. The chapter begins with a discussion of the locational requirements of high technology companies and what community attributes high-tech executives consider important. The chapter then proceeds with a discussion of what Philadelphia, Baltimore, Salt Lake City, Phoenix, and Austin are doing to attract high-tech activities. These urban communities were chosen because they represent the broad spectrum of regional high tech programs that are emerging around the country. Finally, the chapter concludes by drawing observations about the potential benefits of a community based high tech strategy and the problems confronting communities attempting to achieve a high tech growth trajectory for their regions.

HIGH-TECH LOCATION FACTORS

In building a strategy to attract high-tech activities, locational studies that provide information on what factors high tech companies seek when they choose a location are indispensible. A knowledge of locational factors is important to enable a community to determine its comparative advantage in high-tech activities, and what needs to be done to rectify any deficiencies in the community's environment. Because of intense competition, it is important that a community carefully compare high-tech location factors with community attributes, and devise a high-tech strategy accordingly.

The most comprehensive study of high-tech location behavior is the Joint Economic Committee Survey of High Technology Companies in the United States (Premus, 1982). An important contribution of the JEC survey is the finding that high-tech companies are "footloose" in their location choices. In the jargon of location theorists, this means that distance is not an important determinant of where high-tech companies locate. The relatively low ratings given to access to markets and access to raw materials provide evidence for this conclusion (see Tables 2.1 and 2.2).

Being footloose does not mean that high-tech companies will locate at random across the country. Instead it means that they will be drawn to those communities and regions that offer the most attractive locational environment. In this regard, the JEC survey makes another important contribution by revealing what high-tech executives consider to be an attractive locational environment. At the region and community levels, the availability of scientists, engineers, and technicians was listed as the most important locational attribute. Because

TABLE 2.1   Factors that Influence the Regional Location
Choices of High-Technology Companies

Rank	Attribute	Percentage Significant or Very Significant[1]
1	Labor skills/availability	89.3
2	Labor costs	72.2
3	Tax climate within the region	67.2
4	Academic institutions	58.7
5	Cost of living	58.5
6	Transportation	58.4
7	Access to markets	58.1
8	Regional regulatory practices	49.0
9	Energy costs/availability	41.4
10	Cultural amenities	36.8
11	Climate	35.8
12	Access to raw materials	27.6

SOURCE: Joint Economic Committee Survey of High Technology Companies in the United States (Premus, 1982: 23).

1. Respondents were asked to rate each attribute as "very significant, significant, somewhat significant, or no significance" with respect to their location choices. The percentage of very significant and significant responses were added together to obtain an index of overall importance.

of their heavy dependence on R&D inputs, high-tech companies are strongly drawn to the skilled segment of regional labor markets. For policy purposes, it is important to note that workers with technical skills were ranked higher than the more mobile scientists and professional engineers. This suggests that technical colleges and institutions have an important role to play in augmenting the region's supply of technically trained labor, as do advanced centers of education.

The preference to locate near a major university system is another important revelation of the JEC survey. It confirmed the conventional wisdom that a university system confers important benefits on the high-tech community at large. These benefits include educational opportunities for employees and their families, a ready supply of technically trained workers, a source of new ideas from university R&D activities, and cultural and recreational opportunities. While many of these benefits can be obtained by contacts with universities in distant regions, others can be obtained only by physical proximity to the university system.

Other regional and community attributes include the cost of labor, the cost of living, taxes, ample room for expansion, good schools, recreational and cultural opportunities, and an efficient regulatory process, especially in regard to the cost and time of undertaking

**TABLE 2.2   Factors that Influence the Location Choices
of High-Technology Companies Within Regions**

Rank	Attribute	Percentage Significant or Very Significant
1	Availability of workers:	96.1
	Skilled	88.1
	Unskilled	52.4
	Technical	96.1
	Professional	87.3
2	State and/or local government tax structure	85.5
3	Community attitudes toward business	81.9
4	Cost of property and construction	78.8
5	Good transportation for people	76.1
6	Ample area for expansion	75.4
7	Proximity to good schools	70.8
8	Proximity to recreational and cultural opportunities	61.1
9	Good transportation facilities for materials and products	56.9
10	Proximity to customers	46.8
11	Availability of energy supplies	45.6
12	Proximity to raw materials and component supplies	35.7
13	Water supply	35.3
14	Adequate waste treatment facilities	26.4

SOURCE: Joint Economic Committee Survey of High Technology Companies in the United States (Premus, 1982: 25).

business investment and expansion. Finally, community values as reflected in their attitudes toward business are also listed as important in determining the locational environment of a region or community.

The JEC survey is valuable for what it found to be the important locational factors, but the not so important factors should also be recognized. In particular, the JEC found that high-tech executives place little emphasis on the natural environment such as energy, climate, water, waste disposal, or raw materials when choosing a location. It is instructive to note that most of the important locational attributes are determined by the economic, social, and institutional environment within the region.

AGGLOMERATION ECONOMIES

Economists explain the existence of large concentrations of high-tech activities in a few urban regions, notably Route 128, the Silicon Valley, and the Research Triangle Park, in terms of agglomeration advantages (Malecki, 1982). By this they mean that high-tech companies prefer to locate near one another so that they can share a common labor pool, exchange ideas, and obtain specialized resources, such as venture capital funds, consultants, and marketing facilities. Whether or not high-tech companies seek out agglomeration advantages is important to urban communities seeking to attract high-tech activities because agglomeration advantages are one of the most important advantages of urban centers.

The JEC survey asked its respondents to reveal their preferences for an urban environment versus a rural environment. The overwhelming number of respondents, 88 percent, favored an urban location. Over 69 percent of the high-tech executives listed the need for daily interaction with other businesses as an important reason for preferring an urban location. Thus, although the JEC survey did not offer conclusive support for the agglomeration hypothesis, the fact that the majority of respondents preferred an urban environment suggests that urban communities offer a more attractive bundle of locational attributes than rural communities.

SUBURBAN/CENTER CITY LOCATIONS

Where within an urban environment will the high-tech companies locate? The JEC survey did not ask this question, but over 24 percent of the respondents voluntarily revealed a preference for a suburban over a central city location. This indicates that suburban communities with good schools, room for expansion, a skilled labor pool, lower taxes, and recreational and cultural opportunities are the preferred intraurban locations, other things equal.

Before concluding this section, it is useful to point out several observations about the preferred mix of community locational attributes. First, unlike the factors that influence the location of traditional manufacturing companies, such as access to markets, access to raw materials, transportation, and the availability of low-cost labor, state and local governments can exercise considerable control over the locational determinants of high-tech companies. While this observation suggests that high-tech can offer an attractive growth strategy for a local community, it also poses a dilemma. Most of the important high-tech location factors (e.g., education and taxes) are the shared responsibility of the federal, state, and local governments.

Providing the necessary intergovernmental cooperation and coordination may be the largest challenge confronting urban communities attempting to implement a regionwide plan to encourage the emergence of high-tech industries.

A second observation is concerned with the issue of targeting relocating firms with elaborate subsidy schemes. While the JEC survey supports the view that taxes and locational subsidies are important, it suggests that public expenditures for education, training, research, and a good community environment are also important. The upshot of this discussion is that communities would be well advised to focus their development efforts inward on policies that improve the entrepreneurial and innovation climate within their region. This may require additional investments in research and training, but it will also likely require constraints on unnecessary growth in government regulations, spending, and taxes.

see p. 58 for:

HIGH-TECH STRATEGIES

In attempting to steer their economies to a brighter future, many urban communities have in place newly formulated strategies to attract high-tech companies. This section examines a number of these strategies to determine what communities are actually doing to encourage high-tech expansion.

PHILADELPHIA

The City of Philadelphia boasts of the first successful urban research park in the United States (Freedman, 1982). It was an idea that sprang from a 16-acre Science Center, a nonprofit organization owned by 28 universities, colleges, and medical health centers in Pennsylvania (New York Times, April 28, 1982: A20). The University City Science Center, which houses 75 organizations, was created to provide a forum where businesses and universities could combine resources to produce a steady flow of innovation ideas. The University City Science Center project began in 1964 with a grant from the Governor of $500,000, along with an additional $1 million for construction of the facility (New York Times, April 28, 1982: A20). While they were successful in creating innovative ideas, Philadelphia quickly realized that it was missing another main ingredient in building a high-tech environment: an adequate supply of risk capital.

Few venture capital firms are located in Philadelphia and the existing lending institutions are generally quite conservative. Philadelphia's commitment to strengthen the pool of capital originated with a vigorous effort to encourage R&D expansion in order to increase the supply of venture capital and fuel the start-up of firms from the research park. As an inducement, the University of Pennsylvania provided low-cost rental space at the University City Science Center to inspire the business community to engage in firm spin-offs (New York Times, April 28, 1982: A20). To further induce new firms to locate near this research park, the city has committed itself to revitalizing the inner-city area. Thus, innovative products developed at the University City Science Center are expected to fund venture capital activity as new ideas are successfully marketed.

Another aspect of Philadelphia's high-tech strategy is the anticipated use of enterprise zones. In October 1982, the federal government passed legislation giving a tax break for businesses exploring R&D projects. At this time, Philadelphia is considering plans to create an R&D enterprise zone in West Philadelphia's inner-city area (Gruenstein, 1983). The goal of the enterprise zone is to spur a renewed interest in the business community to locate research headquarters facilities in the inner-city area. In conjunction with the enterprise zone strategy, the School of Engineering and Applied Science and the Wharton School have also enacted programs to combine the talents of faculty and students with local business community. These linkages along with enterprise zones are expected to vastly improve the range of incentives for businesses to choose Philadelphia as a headquarters center and provide a portfolio of growing public and private funds in the area.

MARYLAND/BALTIMORE

Maryland does not have a history of a vigorous high-tech entrepreneurial activity, nor is it in a competitive tax position vis-à-vis the Sun Belt areas of the country (Brennan, 1980: 4). What is unique about Maryland is the rapid progress that the public sector is making through the creation of aggressive financial mechanisms. The governor and state legislature have combined forces to make judicious use of state resources to promote active participation of the private sector (Office of Business and Industrial Development: 22). The overall objective of these united efforts is to prompt private investment in human resource development, create capital for technology projects, and most importantly to encourage the migration of existing high-technology firms to the area.

One mechanism the state is employing to foster an improved high-technology climate is a "roundtable" with representatives from high tech companies, financial institutions, public and private universities, and state government. The roundtable committee will strive to improve educational opportunities at all levels of training to promote a larger labor supply to meet the challenges of more technically oriented jobs. Plans to improve the financial climate include the creation of a public venture capital fund and ad hoc teams to encourage the use of financial resources for significant technology projects that will create more jobs and lead to a larger pool of capital (Dealy, 1982: 17-18).

Baltimore has adopted specific measures to improve and emphasize the state's interest in high-tech retention and expansion. The city has created a task force to examine what contribution it can make to improve the quality of the work force and to induce private sector investment in job creation and job training. This includes an educational policy to examine the courses of study that should be available at the primary, secondary, and vocational levels to insure a technically literate work force (Maryland Industrial Development Financing Authority, 1982: 1). A pilot program is also being instituted to train secondary teachers, from different disciplines, in math and science education.

The state legislature has made an effort to create low-cost loans for subdivisions of the state to be developed as industrial property. To fund this endeavor and further the spread of industrial parks, the legislature appropriated $15.6 million in loans from general obligation bonds (Maryland Industrial Development Financing Authority, 1982: 1). Along with this program, the University of Maryland Engineering Research Center has started a technological extension program. This program unites the expertise of professors with the daily engineering problems that smaller companies face as they strive to develop and apply new technology.

UTAH/SALT LAKE CITY

The seed from which Utah's high-technology experience sprouted is the University of Utah. In 1977, the National Science Foundation established funds for four universities to stimulate technological innovation on their campuses. From this program, the University of Utah founded the Utah Innovation Center (Brophy, 1983: 2). The UICI incorporated in 1982 with the help of 20 business and professional people who invested $1 million in start-up capital (Brophy, 1983: 2).

UICI's primary function is to unite new technological findings with entrepreneurs. This process is conducted in the University of Utah Research Park, which provides scientific equipment, a technical library, and several clerical services. UICI funds itself by retaining an equity position or sharing a proprietary interest in the technology developed from R&D projects undertaken on its premises.

Utah's experience is unique because of its focus on spin-off activity. The formation of a company to pursue work begun at the university is only the beginning. For instance, Terra Tek Incorporated spun-off from the university in 1969 as a geoscience firm and has since spawned seven subsidiary firms (Brophy, 1983: 3). Currently, Terra Tek Inc. is continuing to function as an incubator by nurturing new high tech firms. Terra Tek Inc. has been quite successful in its endeavors, with sales increasing from $4.7 million to $12.8 million in six years with over 350 employees. The philosophy behind Terra Tek is that companies are best funded through venture capital, stock sales to employees, and its own profits. When new ideas are generated in the company, they are encouraged to spin-off as a separate entity (Brophy, 1983: 3).

To sidestep the problems of a shortage of local venture capital in Utah, Governor Matheson has promoted the Utah Technology and Innovation Act. Earlier this year the bill passed both houses of the legislature giving authority to the state to create an Economic Development Corporation. The Economic Development Corporation is to be directed by a board of directors composed of two state representatives and six private sector representatives. Among these six members, the president and vice-president are to be selected from the University of Utah Research Park to create a strong university linkage. The goal of this troika of participants is to balance the organization of the board and promote strong local business leadership. Similarly, the Economic Development Corporation is to have triumvirate funding through federal, state, and private sources. Presently, federal funding is being sought at the $2 million level with equal participation expected from the state. To provide a pool of funds for the state to draw upon, the legislature released a portion of Utah's $1 billion pension fund to be used by well-established venture capital firms.

With vigorous support of local businesses, it is expected that government funds will be matched, providing Utah with approximately $20 million in venture capital in the near future. As for the long run, the state anticipates the growing number of firm start-ups will help generate additional capital. This is expected to bring the state into the forefront of high technology with a level of $60 to $80 million in venture capital available to new firms.

ARIZONA/PHOENIX

Phoenix is a forerunner among cities striving for a balanced high-tech program. Their approach consists of state and local business participation. To coordinate the efforts of these different sectors, the governor's office has created an Economic Planning and Development Board and a High Technology Industries Advisory Council for Engineering at Arizona State University (Office of Economic Planning and Development, 1982: 3). In this manner, the governor's office has provided Phoenix with a great deal of information about expected growth of information processing and how the city can meet the requirements of a better trained work force and build a large pool of available capital for investment.

Phoenix's high-tech strategy is designed primarily to attract firms to the area with somewhat less emphasis on spin-off activity. The state's Economic Planning and Development Board directs and coordinates Phoenix's high-tech expansion. The most significant component of their plan was a grant of $32 million to establish a Center for Excellence in Engineering and an additional $13 million given to Arizona State University (ASU) to expand their science program (Center for Research, 1982: 3).

Over the last two years, Arizona has become one of the largest electronics centers in the nation. At ASU, the engineering enrollment has increased at a rate of 20 percent annually (Center for Research, 1982: 2). In order for Phoenix to meet the rising demand for engineers, the governor's office has also created the state's Advisory Council on Community Colleges to forecast job growth in Phoenix and other metropolitan areas, so that the 300 public and private institutions can structure their programs to accommodate for the changing employment demands (Office of Economic Planning and Development, 1982: 12).

The key factor in Phoenix's attempt to create a strong high-tech climate to attract and retain new firms is their Center for Excellence in Engineering. The program is a product of the state's effort to stimulate high-tech activities by augmenting its supply of scientific and technically trained workers. Included in this plan are 68 new faculty positions at ASU and a new 120,000 sq. ft. facility for research. The program calls for $1 million to be made available during the first year and an investment of $31 million over the next four years. This program is funded by state appropriations of $29.5 million for buildings, $8.5 million from the private sector, and an additional $3 million from federal funds for research programs. A primary goal of the program is to repay the investment in a few years by encouraging more engineering firms to relocate or start-up in the state.

A goal of the Center for Excellence in Engineering is to combine the research efforts of different companies in Phoenix with the technical experience of professors located at ASU. The plan for engineering excellence is a product of a comprehensive study by members of the ASU Advisory Council for Engineering and 54 members of six other task forces composed of industry professionals and ASU faculty (Center for Research, 1982: 2). The intent of their combined effort is to create an environment in which technological innovation can thrive.

TEXAS/AUSTIN

Austin is a city with a stylized approach to high-technology growth. In 1982, the city of Austin was selected as the final site for the headquarters of Microelectronics and Computer Technology Corporation (MCC) (Office of the Governor of Texas, 1983: 1). MCC's decision to locate in Austin was largely related to the enormous package of incentives offered by the University of Texas and the governor's office. The plan was developed around the Balcones Research Park of the University of Texas. It provides for a vast expansion of facilities in the research park and the university's microelectronics and computer science program.

MCC is a company owned by 12 American microelectronics and computer companies and controlled by a board of directors representing each of the shareholder companies (MCC Newsletter, 1983b: 1). An advantage to Texas of MCC's decision to locate in Austin is the innovative research that will be made available to companies already located throughout the state.

The MCC proposal was orchestrated by Governor White's office, the city of Austin, the University of Texas, Texas A&M University, and Balcones Research Park to provide an ideal environment for MCC to produce technological innovations. The Balcones Research Park will function as the center for MCC's facility with 20 acres of land leased to the company to house their research (MCC Newsletter, 1983a: 1). At present, the University of Texas Board of Regents has approved $52 million for the construction of six buildings plus utilities, and site preparation (MCC Newsletter, 1983a: 2). Currently, 25 science and engineering laboratories are housed on the campus including Applied Research Laboratories working for the government on R&D contracts, the Center for Earth Science and Engineering, and the Center for Research in Water Resources. Together, these firms will be housed in 950,000 square feet.

The list of amenities offered to MCC is quite extensive and largely focus on the economies derived from close university contact. The plan for MCC begins with a $15 million endowment to create new faculty positions, $5 million to be allocated specifically for new professorships and 6 new endowed chairs, 30 new faculty positions in microelectronics and computer science to be received over a three-year period, $750,000 per year to be available for graduate fellowships, $1 million per year for equipment maintenance and technical personnel and other operating expenses, and 20 acres of land on Balcones Research Park to be leased to MCC for 10 years for a nominal fee with an option to renew the lease (MCC Newsletter, 1983b: 1-2). In addition to the university's support, the state and local communities have arranged the following programs: $20 million in single-family mortgage loans at 200 basis points below the Federal National Mortgage Association's rate, 30-year mortgages without application fees that are subject only to an approved credit application. A commitment of $3 million to provide for any contingencies or gaps in loans for MCC personnel, an Office of Relocation Assistance for placement of MCC personnel spouses, $500,000 to be used to underwrite relocation expenses of MCC personnel such as rental vehicles and other services, and a Lear jet for use of MCC personnel for two years from their decision to locate in Austin (MCC Newsletter, 1983b: 4).

The objective of the plan that Austin made available to MCC was to tailor the city's high-tech strategy to one specific R&D joint venture. This marks a much different approach from other cities opting to arrange a broad list of incentives to attract a variety of firms or to promote firm start-ups. Although Austin's prescription to build a high-technology center does not preclude other types of activity, it is clearly structured to accommodate the needs of the MCC venture. Thus, the key to their success will be the utilization of profits from MCC research to promote new firm start-ups and to attract other high-technology firms to Austin to participate in the research center at Balcones Park.

Several general observations can be gleamed from the high-tech strategies discussed in this section. First, in each case, considerable emphasis is being placed on coordinating the actions of the numerous government and private sector organizations involved in the strategy. Second, the university system is the central focus of most of the strategies. In each case, the communities are attempting to build university linkages to the high-tech business community, by providing training, research, and spin-off activities. Finally, all of the communities are well aware of the limitations of government in creating a spontaneous entrepreneurial climate for high-tech activities. For this reason, they are looking to the private sector to provide leadership in identifying and capitalizing on high-tech opportunities specific to their programs.

SUMMARY AND CONCLUSIONS

The chapter began with an examination of the locational requirements of high-tech companies and proceeded to a discussion of several case studies to see how urban communities are actually attempting to encourage high-tech expansion within their regions. Virtually
all communities have a role in helping their citizens to participate in a
more technologically oriented society, and in helping businesses
within their community to adopt and adjust to the new technologies.
How a community responds to the challenge and to the opportunities
for economic development resulting from technological change is
considered in this chapter to be that community's high-tech strategy.

Many communities around the country are attempting to make
their regions "seedbeds" of technological innovation. While many
observers take a dim view of these efforts, the high-tech strategies
examined in this chapter suggest that this pessimism may be unwar-
ranted. The communities in this nonscientific survey were found to be
promoting their regions as high-tech centers by favoring policies that
stimulate research, entrepreneurship, and technological innovation.
As part of the strategy, they are promoting excellence in education,
venture capital, risk taking, basic research, and ways to facilitate the
transfer of technology from the university to the areas businesses.
Perhaps most important, these communities are attempting to expand
the role of the university in regional economic development by remov-
ing traditional barriers between government, industry, and academia.
To the extent that these community high-tech strategies are reflective
of what is happening in other communities, the nation stands to gain
as more and more communities attempt to capitalize on their role as
incubators of the nation's technological innovations.

Overcoming the technical, financial, and institutional barriers to
technological innovation can be formidable, but for those communities that succeed, the benefits can be substantial. A thriving
high-tech community within a region can be expected to benefit the
region in a number of ways. First, the experiences of Massachusetts
and California have proven that high-tech industries are an important
source of new job growth for regions and the nation. Second, the
presence of high-tech firms within the regions can facilitate the exchange of ideas and the transfer of technology among nontechnology
producing firms throughout the region. To the extent that this occurs,
the competitive structure of the region's economy will be
strengthened, leading to important downstream job and employment
growth effects. The dynamic effects of technological innovation on
job growth are often overlooked in the current discussions over the

job generating potential of high-tech industries. Third, the presence of a large and expanding pool of R&D professionals can act as the "eyes and ears" of the region to important technological developments occurring throughout the nation and the world. The improved information flow can make it much easier for local companies to adjust and respond to economic opportunities and challenges posed as a result of technological change.

In conclusion, for those economic development planners who believe that technological innovation offers an alternative growth path for their community, working with the regions scientific and business communities to formulate a high-tech strategy in the form of a "vision of the future" may be the easiest task. To become an effective instrument of national revitalization, the high-tech strategy must also meet the following challenges:

- — finding adequate sources of funding to support a well conceived, realistic high-tech strategy for the region over a sustained period of time;

- — overcoming the traditional barriers between universities and the business community in order to create an innovative environment beneficial to both parties and the region;

- — finding a way to effectively integrate and coordinate state and local government efforts to attract high-tech development. (Interjurisdictional rivalry is intense in most regions and difficult to surmount; but, an improvement in the timely delivery of services and a more sensible approach to tax policy and regulations would give a region a considerable edge over its rivals.); and

- — creating a venture capital pool to overcome deficiencies in a local business development financing without directly competing with traditional venture capital lenders.

These challenges may cause some communities to turn away from high-tech development but to do so without a careful consideration of the alternatives could be a national loss. After all, technological change is simply getting more from the same resources as a result of a more intelligent use of the resources.

REFERENCES

BRENNAN, P.J. (1980) "Technology of Maryland from clipper ship to satelite." Scientific American (December): 4.

BROPHY, J.J. (1983) A Thriving Partnership: The University and High Tech Industry. Salt Lake City: University of Utah.

Center for Research (1982) Advisory Council of Excellence in Engineering for the Eighties. College of Engineering and Applied Sciences at Arizona State University: 3.

DEALY, J. F. (1982) Report of Governor's Ad Hoc Committee on High Technology submitted to the Honorable Harry Hughes. Annapolis (October).

FREEDMAN, A. E. (1983) "Location factors for technology-based firms: where does Philadelphia stand?" M.A. thesis, Wharton School, University of Pennsylvania.

GRUENSTEIN, J. (1981) "A plan for West Philadelphia." Federal Reserve Bank of Philadelphia (November 13).

HOOVER, E. M. (1975) An Introduction to Regional Economics. New York: Alfred A. Knopf.

Maryland Industrial Development Financing Authority, Committee of Economic and Community Development (1982) Maryland's Umbrella Program: A New Way to Finance Industrial Projects for Small Business in Maryland. (December).

Maryland Department of Economic and Community Development (1982) "Maryland Industrial Land Act." Legal reference Article 41, Sections 438-446, Annotated Code of Maryland (1982 Replacement Volume), as amended.

MCC Newsletter (1983a) "Balcones Research Campus: the University of Texas at Austin." (June).

———(1983b) "Executive summary: the Texas Incentive for Austin." (June).

New York Times (1982) "Science Center helps renewal in Philadelphia." (April 28):

Office of Business and Industrial Development, Maryland Department of Economic and Community Development (1983) Opportunities for Silicon Valley Industry in Maryland. Annapolis.

Office of Economic Planning and Development (1982) Arizona in the Information Age: Opportunities for Growth in the Information Processing Service Industry. (October).

Office of the Governor of Texas (1983) "MCC Austin Incentive Newsletter." (June 13).

PREMUS, R. (1982) Location of High Technology Firms and Regional Development. Washington, DC: Government Printing Office.

3

Cities and the Challenge of the Global Economy

RICHARD V. KNIGHT and GARY GAPPERT

☐ THIS CHAPTER is organized around the assumption that American society is undergoing a major transformation toward an advanced industrial society at the same time that a new global economic system is emerging. One result of these forces is that, in the next 20 years, a new kind of city will develop, a world-class city, a city that will be transnational in many of its transactions.

It is further assumed that it is in both the national and local interest that cities, or city-regions, position themselves in the emerging global economy. Such positioning, ultimately a form of urban strategic planning, requires that cities learn to identify opportunities and to overcome the barriers to their achievement. Since urban growth in the twentieth century has been more "incremental" than "intentional," it may be difficult for local regional communities to develop a new, more strategic, consensus so that they can participate effectively in a competitive world economy.

Other cities might choose to shrink back from participation in the new global economic systems of manufacturing, commerce, trade, and tourism. Some communities have always responded to the challenge of new technologies and new economic opportunities by a process of shrinkage, retreat, and a reassertion of their provincial nature.

In this chapter, we want to identify some of the issues associated with urban development in a changing global economy and to elaborate a new framework by which the economic development of cities can hopefully be better understood. We will end with some speculations about the strategic management of cities in the new, more competitive world economy.

AUTHORS' NOTE: *This material is adapted from "Seven Scenarios of Urban Change" by A. Shostak in G. Gappert and R. Knight,* Cities in the 21st Century *(Beverly Hills, CA: Sage, 1983).*

THE CHANGING GLOBAL CONTEXT[1]

At one level of analysis, it might be suggested that the global economy began in 1492 when Christopher Columbus discovered the New World. At another level international trade patterns can be traced to the establishment of the earliest cities over 6000 years ago. For many European and Oriental cities—such as Venice, Amsterdam, Canton, London, and Paris—the global economy has always been a matter of civic concern. And major multinational companies such as the Hudson Bay and the East India companies in London have formed an important component of their institutional or economic base for several countries.

In the American context, a global city scenario has been readily apparent in only those cities that have served historically as point-of-entry cities (New Orleans, Boston, New York, etc.). Other American cities have more recently taken aggressive steps to insert themselves into the evolving world system of cities (Dallas-Fort Worth, Atlanta, Knoxville, etc.).

Howard U. Perlmutter (1979) has developed and refined for the Philadelphia Tricentennial an international city scenario that might apply to any city. He contends that "in a world which makes 'instantaneous neighboring' technologically possible, the challenge is posed to the city, any city in the world—Do you want to belong to the world?—Do you want to recognize your linkage with the world?—Would you like to establish some roles that you would like to take in the world?" Based upon his experience of years of international living, and his recent part in preparing a scenario for the future of Paris (at the invitation of the French Government), Perlmutter has outlined an international city scenario for Philadelphia.

First, to help guide this scenario, Perlmutter (1979) prepared a definition of the goal:

An international city is a city where ideas, people, and products circulate with relative freedom, where people from other countries are given respect and a sense of dignity, are seen as individuals, and where there is a willingness to bring to that city the best of the world, including what the city itself has, building on the city's own strengths.

Given the city's many ethnic and nationality groups, and their links to nations abroad, Perlmutter is encouraged to think that Philadelphia has a strong foundation for grass roots support of his scenario.

To help get there from here, Perlmutter recommends seven major steps. Philadelphia must immediately work hard to

(1) Develop information systems and export potential: This might take the form of a trading company to locate foreign markets for Philadelphia products and services. Special attention would be paid to ways for small and middle-sized businesses to learn about their world markets.

(2) Build up a regional financial center: This might involve helping local banks begin to act as a network and also seek new foreign branch banking.

(3) Attract foreign direct investments: This might involve wooing foreign business otherwise likely to locate elsewhere in the United States.

(4) Create worldwide recognition: This might take the form of cultural events that build up the city's image and capabilities as a center for world events.

(5) Reorient job training programs to the future: This would seek to match the vocational education process of minorities, unemployed Philadelphians, and young people to the nature of skills that are going to be needed—as these may be known abroad and to global companies.

(6) Protect the "League of Nations" ambience of old neighborhoods: This might mean protecting the status quo—"If the communities are diverse ethnically, allow them to remain diverse; if they are homogeneous and that's the way they want it, then permit them to be homogeneous, still with an international character to the community."

(7) Create an international city hub: The Delaware River area, at Penns' Landing site, is mentioned in this connection. It could include "a world trade center, with hotels, smaller buildings, and apartments, and the Penn's Landing Museum which could become an international exhibit center. This will be a place where the world comes to Philadelphia, a place where Philadelphians can meet the world" (Perlmutter, 1979).

He projects that if even half of these programs were successful, almost 30,000 new jobs would be created.

Such a set of scenario-promoting steps might be easy to envision in Philadelphia, which already receives almost one million foreign visitors a year. Other cities, however, need to come to terms with the steady emergence of a global economy and the pivotal role being played in its development by multinational business conglomerates. These cities can bid for and earn the branch location of dynamic foreign companies, as well as seek international linkages in education, health services, tourism, and trade.

Even more important is the need for the older industrial cities to understand the changing nature of an industrial economy in a global

context. Ever since Daniel Bell introduced his "post-industrial" no-
menclature, people have been befuddled by the notion of a shrinking
industrial sector. The apparent paradox of shrinking blue-collar
employment and expanding industrial production has obscured the
realities of an advanced industrial society where the manufacturing
sector is "served" by those employed in other spheres. The nature of
industrial production in the new global economy must be more
thoroughly understood before policies for urban economic development can be effectively formulated.

A NEW FRAMEWORK

It is likely that there will be a dramatic urban reorientation in the
late 1980s and early 1990s. In the last two decades, purposeful urban
development was often in response to federal urban policy initiatives
and categorical programs replete with "need assessment" requirements that rewarded cities for having relatively "worse" conditions
than elsewhere. In the next two decades, purposeful or intentional
urban development will require a response to the opportunities implicit and inherent in the growth of global markets for both goods and
services.

The context or environment in which economic development
occurs has changed. We are entering a new era. Cities can no longer
afford to deny the realities of the global economy; these realities,
which create opportunities and problems for cities, must be accounted for and accommodated. Future developments must be anticipated so that cities can plan accordingly. Their future hinges upon
their ability to identify new opportunities and to create environments
conducive to such activities. Intentional cities will prosper; accidental cities will continue to decline. National consequences of declining
cities could, however, be minimized by national policies that provide
cities with a framework for making the transition (Knight, 1983).
Transition strategies are needed by cities because, as outlined below,
they have to adjust to the transformation occurring in advanced
industrial society (Hanson, 1983).

As global markets become the norm, and as industries are
rationalized on a worldwide basis, older industrial nations are being
transformed; they are becoming advanced industrial nations. In advanced industrial nations, the transformation involves expanding and
upgrading the more advanced or "knowledge intensive" activities,
while spinning off more traditional manufacturing activities. Con-
comitantly, less-developed countries, by taking on less-skilled and
"de-skilled" operations in traditional manufacturing activities, are

becoming industrialized. Both advanced industrial and developing nations are becoming increasingly integrated into a "common global economy."

Evolution of the global economy can best be conceptualized in terms of the advancement of technology, an increase in international trade, and the international division of labor that results. Adam Smith's adage, that specialization is a function of the size of the market, has taken on a new significance with the advent of instantaneous worldwide communication and "one world" markets. The opportunities and advantages of specialization have increased manyfold over a relatively short period of history.

Advanced industrial activities must utilize state of the art technology in order to remain competitive. Increasingly, technologies are being governed by the organizations that manage them on a worldwide basis. Thus, to understand advanced industrial society, we need to take into consideration how the production and distribution of technology is organized.

A differentiation is occurring in industrialized activities that provides a key to the nature of development of the global economy. Activities that relate directly to the governance of technology tend to be highly centralized and are likely to remain so, but other activities that relate to production, distribution, and sales, which were, at one time, also highly centralized, are becoming increasingly decentralized. Developing countries are increasing their capacity to manufacture goods for export. The rationalization of production activities on a worldwide basis generally involves a shift of manufacturing operations from advanced industrial nations to developing countries (Knight, 1982).

In short, as technology advances, that is, as problem solving becomes more scientific and better managed and as its applications become more widespread, we should anticipate the following: (a) that the international division of labor will increase; that cities in advanced industrial nations will become more and more specialized in activities relating to the governance of technology; (b) that developing countries will continue to expand and upgrade their production capabilities; (c) that international trade will increase; that advanced industrial countries will continue to increase their exports of technology and imports of manufacturing goods; and (d) that developing countries will increase their exports of manufacturing goods and imports of technology. Trade patterns will continue to change as the global economy evolves. Trade reflects comparative advantages that can change over time, but appear to result in an increasingly international division of labor.

CITIES IN THE GLOBAL ECONOMY

Clearly, the evolution of a global economy and the advent of "one world market" has major implications for industrial cities. Their role in the global economy is changing profoundly. More specifically, the functions performed by industrial firms based in these cities are changing. Instead of exporting finished goods, merchandise that requires large numbers of production workers in factories, they are exporting knowledge-intensive, advanced services that require knowledge workers and support staffs in office or laboratory settings.

Advanced industrial cities are now providing technology or industrial know-how for manufacturing operations that dot the globe. Income continues to flow into the economy of advanced industrial cities through industrial corporations, not in payment for goods as it did originally when technology was in its infancy, but in payment for goods as it did originally when technology was in its infancy, but in payment for advanced services that account for an increasing share of the value of finished goods and services. Corporations act as gatekeepers of industrial know-how; they produce some advanced services in house but many are commissioned from specialized technical, managerial, and professional service firms (Knight, 1977, 1980).

The functions performed in advanced industrial cities are changing. And, as the functions performed change, the whole character of the industrial metropolis changes. Communities that began as "mill towns" are evolving into a "world class" of cities. The economies of places that once employed primarily unskilled blue-collar workers are now comprised primarily of highly skilled and white-collar workers. For the most part, the change has been incremental and intergenerational in nature and accommodated in an ad hoc manner. But, in recent years, the situation has changed: Foreign competition has intensified; changes in the economy have been accelerated; overall growth rates in production and employment have declined; and dislocations in the work force have become commonplace. Under these conditions, changes in the social composition and spatial form of the city can become traumatic, indeed. Although some cities have demonstrated great resiliency in times of adversity, most cities need a new framework with which to deal with the challenges posed by the global economy. That framework needs to emphasize human resources.

HUMAN RESOURCES AND
KNOWLEDGE DEVELOPMENT

It will be the human and cultural resources that have been nurtured over the last century or so that will now provide the foundation for the continued development of manufacturing cities. The institutional or man-made attributes of advanced industrial cities will provide a basis for the continued expansion of advanced industrial activities even though traditional manufacturing operations will continue to decline.

The primary force underlying city development in advanced industrial nations today is industrial transformation. Industrial transformation at both the national and global level has its ultimate fallout in cities. The growth of knowledge-intensive activities and the decline of production activities is changing the nature, structure, and form of industrial cities. The foundations for a new type of world city, the advanced industrial city, are being laid as the transformation occurs (Knight, 1984).

Knowledge-intensive activities are becoming the economic backbone of advanced industrial cities. As manufacturing centers evolve into advanced industrial cities, functions related to the production of goods are gradually being replaced by functions related to knowledge-intensive activities, that is, advanced services. In contrast to manufacturing centers, which continue to export finished goods, advanced industrial cities specialize primarily in providing industrial know-how—a broad range of expertise that organizations require if they are to be competitive in global markets.

The transformation from routinized production to knowledge-intensive activities is sufficiently advanced in some places that it is becoming difficult to ignore or deny. Akron, for example, which was established as a mill town and grew into a major manufacturing center, no longer manufactures tires, for which it is world famous. But Akron continues to develop because it has become one of the principal concentrations of industrial know-how that is needed by the growing world tire, rubber, and polymer industries. And the prospects for sustaining development there are good. The demand for the knowledge base of firms headquartered in Akron will continue to increase as long as the industry grows, technological advances are incorporated into worldwide operations, and the leading firms remain based in Akron.

Akron's role has changed: It is now a knowledge center serving global markets. The technologies and companies incubated there now provide the foundation for further development and diversification of knowledge-intensive activities. For example, Goodyear's aerospace research draws on and complements a wide range of technologies that are based in Akron and neighboring cities in northeast Ohio. If anything, the city's development potential has been enhanced as it has evolved; it's development potential certainly has not been diminished by the transformation of its economic base. But it will take time for Akron or any other city whose functions are changing to realize their potential. In order to realize their potential as a world city, they will have to make the transition from accidental growth to intentional development.

The degree to which advanced service functions have already replaced traditional manufacturing functions and the rate at which the transformation to knowledge-intensive activities is occuring varies tremendously in American industrial cities. In some places, the change is welcomed: In others, resistance is strong.

CITIES AND KNOWLEDGE AS A FACTOR OF PRODUCTION

The basic issues concerning city development cannot be fully defined or effectively addressed unless they are considered in an international context and over the long term. National, provincial, and parochial concerns tend to obfuscate issues by focusing attention on symptoms and short-term problems. While there is great concern over the erosion of old values and traditional types of economic opportunity, namely the decline of jobs in manufacturing, there is little concern or consideration given to the new forms of wealth creation, the expansion of opportunity in knowledge-related activities. The battle lines have already been drawn; they are based on old assumptions about how wealth is created. The conventional wisdom supports the chasing of smokestacks (production facilities) and this usually preempts other longer-term strategies such as recruiting talent or winning minds. Even the recent rash of highly touted programs aimed at "high-tech" and "high-touch" industries focus more on routinized manufacturing operations than on the source or knowledge base from which the technology springs. Knowledge-intensive activities are not viewed as being productive because they are not performed on the production line.

The role of knowledge as a primary means of production in advanced industrial cities will not be understood until local industrial activities are viewed in the context of the international operations of which they are a part. But this requires new concepts and a few conceptual framework because knowledge workers are not productive in the traditional sense.

Knowledge activities are usually performed at considerable distance both in time and space from actual production activities. Knowledge is not usually produced by workers on the production line and thus cannot be measured in terms of output per worker per hour. Knowledge is an intangible and its value or the productivity of the knowledge worker has to be measured in terms of the indirect and long-term consequences of such work. Clearly products cannot be manufactured or merchandised without know-how, but how can the contribution of knowledge, vis-á-vis the contribution of the other factors of production, land, labor, and capital, be ascertained?

Such questions are usually brushed aside because they cannot be answered using generally accepted accounting procedures. This is unfortunate because productivity is usually attributed to production workers or capital, which means that indirects or overheads are usually viewed as being unproductive. This historical bias in how wealth creation is perceived is most unfortunate because it can lead to underinvestment in knowledge resources, inordinately high compensation for production workers, low compensation for knowledge workers, and the sale of intellectual properties such as patents. These issues will become increasingly important as society becomes more knowledge based.

One of the main contributions of knowledge workers is to create wealth through increasing the efficiency in all steps of the production and distribution process. But productivity gains are usually attributed to production workers rather than to the knowledge workers that thought up the improvements. Increases in production per worker hour are rarely attributed to their real source such as research and development, engineering advances, value engineering, quality control, management or professional services, and so on. They are quite arbitrarily attributed to the person closest to the actual production or sale of the final product.

This bias in "value added" and productivity measurements needs to be corrected. Errors of measurement were inconsequential when the knowledge content of goods and services was small and when knowledge and production were located at the same site. Now, however, the knowledge content is highly significant and, with the international division of labor and operations spread around the world, linkages between different steps in production need to be specified and accounted for. We must, therefore, revise our methodology and begin attributing value to its source, so that knowledge workers will be fairly compensated. How can we have confidence in advanced industrial cities if we continue to think in terms of traditional production activities, if we continue to insist that lawyers, accountants, bankers, teachers, and managers are unproductive, or if we think only in terms of the qualities of tangible products that can be measured, weighed, or counted?

New forms of wealth that are being created in advanced industrial society are knowledge-intensive: They require a lot of know-how, technology, and information. Some types of knowledge take the form of intellectual properties such as patents, copyrights, plans, contracts, degrees, and titles, which are negotiable, but it is probably reasonable to argue that most knowledge exists primarily in the minds and memories of knowledge workers and cannot be treated as property in the same way as commodities are, for example.

Knowledge workers, working individually or collectively as members of organizations, are able to access and apply knowledge through elaborate networks that are both formalized and informal in nature. Knowledge workers create wealth by developing, implementing, and managing ideas; their primary activities are thinking, learning, and communicating; their basic material is information, which they gather, process, and distribute. The role of information is only a small part of the process, it tends to be overplayed because, again, it can be measured. Although information regarding technology can be symbolized and thus easily communicated and widely disseminated, the knowledge base on which the technology is founded becomes increasingly complex and difficult to access. The knowledge base cannot be replicated: It must be passed on to the succeeding generations on a one-to-one basis. This is why knowledge is concentrated in cities and why, historically, the primary function of cities has been the forging and transferrence of cultural values.

Knowledge-intensive activities need to be defined in operational terms; they involve working with knowledge, technology, or information such as applying, servicing, and advancing technology, adapting technology to new situations, to new problems, and to new markets, merging new and old technologies, and expanding these activities around the world. Production and distribution of goods and services is becoming increasingly knowledge intensive or "science based," and, as already mentioned, operations are being rationalized on a worldwide basis.

THE ROLE OF CITIES

As this transformation of advanced industrial society proceeds, the role of cities in regional, national, and in the international economy and the context in which they develop changes significantly. New development processes, new socioeconomic forces and constraints are acting upon cities, causing their development to take new forms. New types of world cities are being established. As shown in Table 1, roughly 20 cities in the world are the respective homes of two

TABLE 3.1   Cities with More Than One Corporation
with $6 Million in Sales

City	Number
Tokyo, Japan	21
London, UK	17
Paris, France	15
Osaka, Japan	11
New York	9
Montreal, Canada	4
Pittsburgh, PA	4
Rome, Italy	3
Frankfurt, Germany	3
Essen, Germany	3
Los Angeles, CA	3
Dusseldorf, Germany	2
Madrid, Spain	2
Tel Aviv, Israel	2
Toronto, Canada	2
Netherlands/UK	2
Chicago, IL	2
St. Louis, MO	2

SOURCE: *Forbes*, July 4, 1983.

or more corporations that achieve at least $6 million worth of sales on an annual basis. Another 66 cities in the world are each the home of at least one corporation with worldwide sales over $6 million and include Peoria, Caracus, Turin, Bombay, and Seoul. Smaller urban places everywhere are able to serve a multinational corporation.

Structural changes in the national economy can be traumatic at the regional and local level or for specific industries, occupations, or segments of the labor force. Traditional goods production activities are declining along a wide front in the older manufacturing regions while advanced industrial and other knowledge-intensive activities expand selectively in certain types of communities and require special types of work skills; some difficult social and economic adjustments have to be accommodated.

The development process is very uneven. Institutional linkages between leading and lagging areas are critical and need to be maintained so that continued national development is not jeopardized by the adjustment problems of special interest groups. Gains and losses and their associated costs and benefits fall on different populations, on different communities, and often on different generations and in different time periods. Structural changes that are not well planned can give rise to economic and social disparities between areas that will, if given time, slow and possibly threaten the whole development process unless they are ameliorated.

Social and economic disparities arise because areas at the leading edges prosper as knowledge intensive activities expand, while areas built in an earlier era for unskilled factory workers become distressed. Once disparities between regions or between inner-city neighborhoods and suburbs or between population groups become established, they are difficult to arrest or diminish. In fact, as disparities widen, they are increasingly factored into all types of decisions, which further polarizes the situation. The consequences of neglect cannot be ignored in the long run. History is replete with examples of societies that slowly disintegrated or were violently collapsed by disenfranchised groups.

Structural changes and disparities occur at all levels of organization—from neighborhood blocks to nations—thus the responsibility for managing change has to be shared, distributed, and coordinated throughout the national hierarchy of urban places. Disparities are usually greatest and most difficult to correct in cities, regions, or nations where the structural changes are most pronounced. Industrial cities established early in the Industrial Revolution, especially those established around "basic industries," are probably the most vulnerable to the adverse impacts of the global economy. In the main, these early industrial settlements were poorly designed and built at the offset. These accidental cities face the greatest challenge.

Cities now need to shift from reactive to proactive policies so that they can position themselves in the rapidly evolving global economy. In order to do this, cities will have to be able to differentiate between activities in which they can compete effectively in world markets and those in which they have outgrown or lost their competitive edge. Cities will have to be able to anticipate and then be prepared to accommodate structural changes induced by the global economy. Cities will have to be able to identify and then to build on their strengths in ways that will enhance their development. Not many places now have these capabilities and to learn strategic management.

The success of strategic initiatives at the city-region level have major national implications. Unless cities in the United States are able to identify and capture these new opportunities and accommodate structural changes more effectively, the nation will lose its leading edge in the global economy. It is highly unlikely, however, that city-regions will be able to sufficiently structure their efforts to make the transition from accidental growth to intentional development expeditiously unless it becomes a high national priority.

A national cities policy that advocates city-region development and supports local initiatives is urgently needed. A national cities policy would have to provide both a framework and incentives so that all city-regions will have an opportunity to make the transition to intentional development, a transition that is required if they are to sustain their development. Cities are the basic building blocks of advanced industrial society; they are on the cutting edge of the global economy. As our cities fare, so will the nation.

STRATEGIC MANAGEMENT AND NATIONAL POLICY

In the late 1980s, the United States is likely to develop a new set of policy priorities. These priorities are likely to reflect (1) a concern with a *global strategy,* (2) some form of coherent *industrial policy,* (3) incentives for *city-region* cooperation, and (4) an emphasis on innovations and *strategic management.*

New national policy debates are already developing at the creative margins of the "conventional wisdom." These new policy issues are not yet part of what the SRI futurists call the Official Future but their development has begun. For instance, the geopolitics of the next half century are not yet apparent but the concern for a global strategy is emerging rapidly. Congressman Mickey Edwards of Oklahoma has written:

> There is a desperate need for a truly comprehensive global strategy that is broadly focused and includes all aspects of the military, economic, industrial humanitarian and psychological tools at our disposal.

He goes on to state:

> The emergence of a technological society, and the disappearance of smokestack industry, is neither a military factor nor a diplomatic factor, but it is an essential part of the equation in developing a viable global strategy whether we protect our industrial capacity with tariffs or develop new means of ensuring the availability of outside sources.

A global strategy must concern itself with different urban scenarios such as those suggested by Shostak (1982) for Philadelphia. And different global scenarios featuring both catastrophes (boat people, mininuclear wars, etc.) and achievements (solar power efficiency, third world consumers, etc.) must be developed with an eye for their urban consequences.

The public concern with a national industrial policy has preceded the new interest in a global strategy but is complementary to it. The new commission on industrial competitiveness is one reflection of that concern. Some cities claim that we already have a mare's nest of industrial policies that are contradictory and confusing. Recent hearings by the Joint Economic Committee have explored in depth the concept and realities of industrial policies. The primary objective of such policies is to introduce a major coordination and planning role for some kind of federal presence. Critics contend that such a policy would be going against the natural shifts and adjustments toward decentralization that are now occuring. Instead, an industrial policy would best emphasize federal incentives for research and development and educational excellence.

In terms of strategic planning the new concerns for both a global strategy and an industrial policy are indicators of environmental turbulence for institutions in both the public and private sectors. The global and national environments are full of new uncertainties and new forms of city-regional cooperation may be necessary in order for local institutions to take advantage of the new opportunities inherent in an expanding global economy.

The real challenge for advanced industrial nations lies at the city-region level. The challenge is one of sustaining development in the context of the evolving global economy. City-regions must be able to play a key role in facilitating the transformation of industrial society. It is at this level that opportunities in the global economy can be identified and captured. Moreover, cities need to understand the nature of the global forces that underlie structural changes in their particular area so that they can govern them more effectively. Denying and resisting such changes is dysfunctional: It contributes to the dissipation of the scarce local resources and undermines confidence in their economic base and in the future of their city.

The concern of a city-region is certainly not new. It was first developed by Dickinson (1947) in his book on urban regionalism. Friedmann (1956) has also elaborated the planning significance of city-dominated functional regions.

In his more recent (1979) book, *Territory and Function,* Friedmann has elaborated the need for a paradigm shift in regional planning in a global economy. He cites two institutional innovations that stand out as useful mechanisms to deal with what he calls the "problems of residual areas." These are (a) the regional development authority and (b) the regional development bank.

Although Friedmann has been primarily concerned with the gradual elimination of the "periphery" on a national scale by substituting for it a single independent system of urban regions, the new concerns for growth management on a global scale have shifted the concern to the strategic chores facing developed cities already linked into the emerging global markets. He urges that attention be paid to "The development of the bases of communal wealth, land and water, good health, knowledge, and skills."

The strategic planning problem though for city-regions, however, remains the lack of incentives for collaborative strategies on a regional scale. The North Texas Commission and the Metroplex concept in the Dallas-Fort Worth area remains perhaps a solitary example of the value of such approaches. Unigov for Indianopolis and the tax sharing arrangements in Minneapolis-St. Paul could evolve into more strategic direction.

The final new policy priority for cities facing the new realities of a global economy is to adapt new strategic management policies and techniques to foster a better climate for innovations in all sectors of their communities.

There is a need perhaps for a unique marriage between corporate strategies for innovation with the environmental movement's concerns for regional self-reliance and, to some extent, self-sufficiency. As discussed elsewhere (Gappert, 1979) a strategic reconstruction perspective is necessary. This strategic perspective must be concerned with the qualities of institutional vitality necessary to plan cities of intentional change, as well as to identify more precisely the nature of new opportunities generated by new technologies and the global economy. The successful and superior cities of the late twentieth and early twenty-first century will be those that have developed institutions that can (1) manage knowledge, (2) promote innovation, and (3) guide scarce resources toward strategic priorities. Such cities are likely to have strong external links to significant economic and political networks as well as good interorganizational links at several levels within the local region.

REFERENCES

DICKINSON, R.E. (1947) City Region and Regionalism. London: Kegan Paul Trench.

FRIEDMANN, J. (1956) "The concept of a planning region." Land Economics 32: 1-13.

————with C. WEAVER (1979) Territory and Function: The Education of Regional Planning. Los Angeles: University of California Press.

GAPPERT, G. (1979) Post Affluent America: The Social Economy of the Future. New York: Franklin Watts.

HANSON, R. [ed.] (1983) Rethinking Urban Policy: Urban Development in an Advanced Economy. Washington, DC: National Academy Press.

KNIGHT, R. V. (1984) "The advanced industrial city: a new type of world city." Presented at the Conference on the Future of the Metropolis. Technische Universitat Berlin, October 26.

————(1983) "Towards a national cities policy." Prepared for the Committee on National Urban Policy, National Research Council, Washington, DC.

————(1982) "City development in advanced industrial societies," Chapter 3 in G. Gappert and R. V. Knight (eds.) Cities in the 21st Century. Beverly Hills, CA: Sage.

————(1980) "The region's economy: transition to what?" Cleveland: Cleveland State University.

————(1977) "The Cleveland economy in transition: implications for the future." Cleveland: Cleveland State University.

PERLMUTTER, H. (1979) "Philadelphia: the emerging international city." Philadelphia: LaSalle College.

SHOSTAK, A. (1982) "Seven scenarios of urban change," in G. Gappert and R. V. Knight (eds.) Cities in the 21st Century. Beverly Hills, CA: Sage.

4

Business Formation and Investment in the Minority Community

GAVIN M. CHEN

☐ CENSUS DATA since the early 1900s have consistently shown that the minority population is primarily urban. Employment opportunities have served as a magnet attracting rural minorities and immigrants to urban areas. However, the demand and supply of jobs and the labor force size have not usually been equated either in terms of numbers or skill. Thus, urban areas have high unemployment rates and other concomitant ills. The minority urban community especially suffers from high unemployment, crime, and low quality of life.

Urban development strategies have often focused on social programs to upgrade life quality while ignoring the economic aspects of causes. Since minority communities are usually hard hit by urban ills, urban development programs should address the economic issues of jobs and revenue flows as a salve to urban ailments. Expansion of present firm capacity cannot meet job demands because additional economies of scale are not significant. Therefore, additional and new job opportunities are needed. This is why entrepreneurship is of strategic importance.

A viable strategy for urban development should not only include economic growth through business development but must also encourage minority entrepreneurship. Minority firms typically employ a largely minority labor force; the results often include the positive community development aspects of institution building and an enhanced minority political power base. This chapter will address the stimulus to minority entrepreneurship, the general perspective on investing in minority businesses, and will pose some policy guidelines for the future.

AUTHOR'S NOTE: *The views expressed in this chapter are those of the author alone and do not necessarily represent the views of the Minority Business Development Agency or the U.S. Department of Commerce.*

BUSINESS FORMATION

Business formation is probably the singlemost important contributor to the economic development process (Baumol, 1980). Established firms do not grow in employment at the rate needed to support economic growth. This growth comes about through firm births. Firm births provide the economy with new and additional employment opportunities. A new increase in firms and employment over time results in growth. Unfortunately, the minority community has lagged significantly behind the nonminority community in business formation (Kirchoff et al., 1982). Minority business formation will aid in reducing high minority unemployment rates because minority businesses primarily use a minority labor force.

WHAT MAKES A MINORITY ENTREPRENEUR?

The decision to start and operate a business is a choice governed by an individual's perception of income, employment and status, training and education, business risk and opportunity, role model influence, and capital availability. Census data in 1977 show a minority business participation rate one-fifth that for nonminority business (Kirchhoff et al., 1982). This could mean that either minority businesses have a lower start rate or they have a higher failure rate. Both are true. Why is there a lower start rate?

Entrepreneurship is important to the development process; but, how do you develop entrepreneurship? There are a number of institutes concerned primarily with this issue (USDOC/SBA, 1981). They evaluate the individual and then determine whether the characteristics of that individual match their assessment of the needs and requirements of an entrepreneur. The problem with these studies is that in attempting to quantitatively predict individual behavior, an aggregate rating scale is used. With aggregation, the nuances and variables apparent from the individual choice perspective are lost. These individual traits are classified in the literature as need achievement, role model influence, and social mobility and status.

NEED ACHIEVEMENT

The idea of need achievement was studied by McClelland (1980; 1969); however, the subjectivity of his defined variables and the measurement problems detracted from his theory.

Briefly, entrepreneurship motivated by need achievement occurs when the decision to start a business is made because the individual wants to be independent, seeks excellence for excellence's sake, feels that his or her destiny is inner-directed, and needs some type of acceptance feedback. (Social or peer group acceptance implicitly assumes that entrepreneurship is viewed as an independent, income-producing career choice.) McClelland (1969) measured need achievement variables: (1) a moderate risk preference; (2) inner-directed influence; (3) an inner assurance of success; (4) a need for feedback of acceptance; (5) the capacity to foresee and plan for the unexpected; and (6) an interest in excellence for the sake of excellence.

McClelland developed his idea of need achievement motivation in an attempt to predict entrepreneurship. Bearse (1982) refuted this attempt on the grounds that studies of entrepreneurship only look at the postchoice scenario and not the environment at the time of choice. Nevertheless, by aggregating McClelland's specific theory of need achievement with the general theory of individual choice, we can obtain an explanation of the behavior pattern for the majority population.

McClelland explains the low Minority Business Enterprises (MBE) participation rate by saying that nonwhites have a lower need achievement motivation than whites because they do not possess the inner assurance of success, a derivation from slavery and discrimination (Bates, 1973). Furthermore, the perception of entrepreneurship is not the same because it is not the symbol of independence nor economic self-sufficiency in the black community as it is in the white community. This difference could also occur because minorities do not view entrepreneurship as a career choice since there is an obvious trade-off with an opportunity cost from employment (Brimmer and Terrell, 1971; and Morse, 1979).

Light (1972) and Sowell (1982) concurred, stating that first and second generation minority immigrants (Asians and West Indians) have high entrepreneurship rates because they have not been confronted with extended overt and covert discrimination. This assertion is similar to Brockhaus's (1981) "displaced person theory." Light (1972) further added that Asians have a high participation rate because of the perception among Asians that entrepreneurship is a vehicle for independence and social mobility. This idea is borne out by Gamble (1982) who surveyed California minority entrepreneurs and concluded that minorities patronized MBEs, had a positive image of the MBEs' worth, and contribution to their community.

SOCIAL STATUS AND MOBILITY

In the general community, the entrepreneur has social prominence. Entrepreneurship is also viewed as providing independence and economic self-sufficiency and is thus a vehicle for social mobility. Business leaders are also community leaders.

Coles (1956) and Caplovitz (1973) asserted that minorities view entrepreneurship as providing independence but not social standing. Blackwell (1974) went further by stating that the professions of law, medicine, and dentistry provide not only economic returns but are more socially acceptable career choices than entrepreneurship among blacks. Furthermore, because of a history of occupational choice restrictions, education and the resultant employment opportunities were considered better vehicles for economic self-sufficiency and community status. In another study, Brimmer and Terrell (1971) asserted that minority business development is economically inefficient and that corporate employment provides better economic returns and social standing.

This theory partially explains the differential between the minority and nonminority general business participation rate. However, a question other than that of sociocultural value arises. This explanation places the entire choice selection within the realm of social feedback and acceptance and ignores the effects of individual experiences and desires; regardless of how they were shaped by society, education, and environment.

By asserting that entrepreneurship does not provide as appealing a career choice as other professions, Caplovitz relegates the entrepreneurship choice and opportunity to the lower socioeconomic stratum of the minority community. Unfortunately, this stratum is constrained because of a lack of advanced educational opportunities, capital limitations, general unawareness of business opportunities, and absence of role models. Entrepreneurial choice is a factor associated in the nonminority community with social status, economic independence, and self-sufficiency or permanent income. In the minority community, this is not the case because of the size of MBEs, the economic attractiveness of alternative career choices, and the social importance of other choices. This difference in perception supposedly explains the apparent difference in participation rates between the minority and nonminority population.

The importance of role models in the entrepreneurial decision process is documented throughout the literature, especially by Sexton (1980) who stated that the decision is significantly influenced by association with other entrepreneurs, especially if those entrepreneurs were parents. The idea of the role model is that if an individual is familiar with an entrepreneur, this role model will influence that individual's career choice toward entrepreneurship. Sexton (1981) asserts that this is a important factor and agrees with Caplovitz (1973) that this relationship could very well be the cause of the participation rate differential between minorities and non-minorities.

Bates (1973) agreed with that assessment, but stated that *there is* a minority business tradition although it is perverse. That is, there has always been a minority business sector, only that it is limited and restricted by industry and customer appeal so that the role model it provides is a negative one. It seems that the presence of a role model is essential to the entrepreneurship decision; however, does it explain the relatively low minority participation rate?

In another study, Brockhaus (1981) pointed out that other variables might override the immediacy of the decision. Although the role model theory is prevalent, other factors (for example, employment experience or employment opportunities and choice) could influence the timing of the entrepreneurship decision (capital, basic needs/gratification, etc.).

LITERATURE SUMMARY

Of the three general theories proposed as explanation for the disparity between minority and nonminority business participation, the role model theory comes closest to completely explaining that differential although all the theories provide some insights. Essentially, the entrepreneurship decision is an individual decision and the combination of factors influencing that decision is never the same for different individuals.

From the preceding literature survey it can be concluded that (1) there is a minority/nonminority participation rate differential caused by a combination of factors such as the dearth of role models and the availability and attractiveness of other career opportunities; and (2) that MBE owners are older, more educated, and as a result have a longer employment experience than their nonminority counterparts.

The second postulate in part explains the participation differential because the minority population is on the whole younger, less educated, and more prone to higher unemployment rates and longer periods of unemployment than the nonminority population. This, in turn, is primarily a function of discrimination—because if both minority and nonminority individuals are exposed to similar econocultural experiences, the minority participation rate will still be lower because of external factors such as discrimination, capital availability, lack of information, and so forth.

Additionally, from the preceding literature, survey certain factors are apparent. First, it is difficult to prejudge or identify entrepreneur-

ial skills in an individual. Tests developed for entrepreneurial character trait assessments are statistical, postentrepreneurial choice measures that are not very useful because of significant individual behavior
aberrations. Second, the role model is an important factor in the
entrepreneurial career choice decision.

Third, the minority individual also faces capital constraints that
delay the entrepreneurial choice until the second or third career
choice stages are reached (Ronstadt, 1982). This means that, on the
whole, the minority entrepreneur will be older and more experienced
when the choice is made.

Three hypotheses emerge from the literature review:

(1) An older and more experienced minority entrepreneur may have
    delayed career choice.
(2) A higher proportion of minority entrepreneurs exists in smaller
    urban areas where minority residential/employment is more con-
    centrated, such as South-East U.S.A. (role model).
(3) A high degree of relationship exists between the entrepreneur's
    education and age given a business type.

SOME EMPIRICAL FINDINGS

This section describes some empirical findings based upon a data
set of 166 observations representing a national sample of the clients of
the Minority Business Development Agency's business development
network. A correlation matrix of eleven variables was prepared. The
data were then broken down by region, by education, and business
types. The variables are business type (BTYPE), business age
(BAGE), owner's age (OAGE), owner's education (OED), owner's
sex (SEX), ethinicity (ETHNIC), region (REG), urban location
(URBAN), total population (TPOP), minority population (MPOP),
and minority business entreprises (MEE). The statistically significant
(.10 level) correlation coefficients are shown in Table 4.1. Tables 4.2
and 4.3 show significant correlations by industry.

The proxy measure of the effect of the role model on minority
entrepreneurship is the degree and significance of the correlations
between the total population and the minority population and the
minority population and the number of minority businesses. To test
hypothesis two above, the variables of region and urban location were
added to total population, minority population, and number of MBEs.
While the absence of a high and significant correlation coefficient
among any of these variables does not imply that hypothesis is in-
valid, all variables would have to be significantly correlated to affirm
that hypothesis. So hypothesis two is unsubstantiated.

TABLE 4.1 Correlation Matrix

	BTYPE	BAGE	OAGE	OSEX	ETHNIC	OED	REGION	URBAN	TROP	NPOP	MBE
BTYPE	–	–	–	0.15768**	–	–	–	–	–	-0.13440*	-0.18924**
BAGE		–	-0.14341*	–	–	-0.20257***	0.14807*	–	–	–	–
OAGE			–	–	–	–	–	–	–	–	–
OSEX				–	–	0.14893*	–	–	–	–	–
ETHNIC					–	–	0.31291***	0.37566***	0.21900***	0.24945***	0.22424***
OED						–	–	–	–	–	–
REGION							–	0.29279***	0.16076**	0.24130***	0.21372***
URBAN								–	-0.21030***	-0.17838**	-0.17016***
TPOP									–	0.95324***	–
MPOP										0.84223***	0.94798***
MBE											–

*Significant at the .10 level.
**Significant at the .05 level.
***Significant at the .01 level.

TABLE 4.2   MBE Characteristics: Overall and by Business Type

Business Type	Variables (Years)		
	BAGE	OAGE	OED
Overall	8.1	35.2	Vocation/trade/some college
Agriculture	2-5	31-5	College graduate
Construction	4-62	37-04	Trade/vocational/some college
Manufacturing	9-33	43-7	High school
Transportation & public utilities	9-33	43-7	High school
Retail trade	7-15	34-2	High school
Wholesale trade	5-6	35-1	High school
Finance, insurance real estate	2-6	33	Graduate/professional school
Services	7-7	34-9	Trade/vocational/some college

The relationship between business types and business age and minority population and number of MBEs seem consistent with theoretical explanations. The newer the business, the higher the business types.[1] High numbered business types (from definition) are more easy to enter but might require education or formal training. The combination of urban concentration (Urban/MPOP, MBE/Urban, TPOP/MPOP) and the relationship between MPOP/MBE implies the significance of the role model in most instances (BAGE/MBE). This is particularly interesting in the construction industry where the younger the minority entrepreneur, the higher the educational attainment. This indicates that minorities are opting for more technical and formal education and are using this more formal training in lieu of apprenticeship and long employment prior to the entrepreneurial career choice.

The relationship defined by the population variables, the urban variable, and the number of business enterprises is one of black concentration and an increase in the number of black businesses. This could indicate a role model influence in the Retail Trade Industry.[2] Although the above implies the presence of a role model influence on the MBE entrepreneurial decision, evidence is not sufficient to positively make such a statement.

TABLE 4.3 MBE Correlation Coefficients by Industry

Industry	OAGE/ OED	SEX/ OED	OAGE/ MPOP	OAGE/ MBE	URBAN/ ETHNIC	REGION/ MPOP	URBAN/ TPOP	TPOP/ MPOP	TPOP/ MBE	MPOP/ MBE
					Variables					
Construction	−50.25									99.6
Manufacturing						47.6				
Transportation & public utilities										98.61
Retail trade			28.53	29.14	54.4		−28.28	96.92	85.4	93.7
Wholesale trade								99.9	91.7	92.7
Services		37.9			39.7			98.5	92.4	91.7

Generally, it can be seen that minority entrepreneurship is a male-dominated career choice and the minority entrepreneur is generally older and possesses some formal educational training beyond the high school level although somewhat lower than that of non-minorities. We can hypothesize that capital scarcity is one of the reasons for the entrepreneurial choice delay. Additionally, the information analyzed does not lend itself to strong positive statement on the influence of the role model, nor does it support the converse of the argument.

The data further show that nonblack minority entrepreneurs are located in smaller urban areas and are more likely to have educational attainment above the high school level. Unfortunately, the data do not contain information on employment experience nor formal training type so it could not be discerned whether the delay in the entrepreneurial career choice was planned or the choice was made because of external circumstances.

The regional breakdown did not substantiate the role model argument as hypothesized, although the breakdown by industrial type did imply some role model influence. This could be indicative of the fact that the specified role model proxy variables are incorrect, or that entrepreneurial role model is more industry specific or enterprise specific than merely being a general entrepreneurial role model.

Finally, I expected to see a general relationship between age and education for each given business type. This was relevant in the construction industry where the data showed that young minority entrepreneurs were substituting formal education for apprenticeship or on-the-job training. This bears out the assertion that younger entrepreneurs invested more in human capital development for income, social status, and social mobility reasons. Because of the lumping together of the service industry types, this analysis could not be performed for the more technically demanding areas such as health, professional, and business services.

The minority manufacturing entrepreneur is much older: 44 years, which fits Ronstadt's third career choice stage. The probability is that this individual was previously employed in relevant capacities for some time thus developing on-the-job expertise. However, even though the career choice was delayed, there was no compensating increase in formal education as this group possessed only a formal high school education. Again, data inadequacies precluded further analysis on the formal education/job experience variables that would undoubtedly have provided a wealth of information.

BUSINESS INVESTMENT

Capital availability for minority business entreprises has always been a problem. This need has stunted the general growth and development of that business sector. The federal government, in recognizing this problem, implemented financial efforts to alleviate the problem by providing for both debt (Small Business Administration [SBA]) and equity (MESBIC) financial instruments. Still, both minority and small businesses have been significantly excluded from private capital markets.[3] The general investment consensus is that there is a certain degree of risk associated with minority business investment, a degree of risk perception that offsets the expected returns. An adequate capital pool will be available only if minority businesses are perceived as financially solvent with a certain degree of longevity and the expected returns meet investor utility functions.

FINANCIAL RATIOS

This section will perform comparative ratio analysis and observe the foreclosure probability of minority businesses vis-à-vis the general economy. In so doing, it shows that minority businesses are viable alternatives as investment choices.

For purpose of discussion, the ratio analysis will be categorized so that certain ratios will measure financial solvency and others will measure profitability. Table 4.4 shows the ratios used in the analysis and their composition.

The data analyzed do not follow normal distribution because they are drawn from a controlled sample. This makes the data suspect as a definitive base; however, the intent here is to provide an argumentative base for rational investor choices and that the minority business sector should be an investment alternative. Tables 4.5 and 4.6 show the investment ratios by business type for minority and nonminority businesses.

Service industry. The control group seems somewhat stronger in terms of the short-term financial capacity as evidenced in the higher quick and current ratio and lower liabilities to net worth ratio although minority businesses do have a lower average collection period and higher inventory replacement ratio.

In terms of efficiency and profitability, the minority group seems more efficient although the profitability picture is mixed. In short, the minority groups in the service industry area do not seem like a good alternative.

Retail trade. The minority group has a higher quick and current ratio while the control group has a lower financial leverage factor and

**TABLE 4.4   Financial Solvency and Operating
Efficiency/Profitability Ratios**

Financial Solvency Ratios

These ratios measure the firm's ability to meet short-term/current debt obligations
requiring immediate cash payouts.

Current	current assets: current debts
Quick	cash availability to meet immediate cash demand
Collection	the average length of the collection time from invoicing to collection of payment
Inventory	the average length of time that saleable assets are processed and held until converted to revenue
Total Liabilities	the ability of the firm to meet both short-term and long-term debts

Operating Efficiency and Profitability

The ratios measure the firm's return investment and the efficiency of its operations

Collection	defined above, this ratio implies asset management
Inventory	again defined above and also measuring asset management
Return on Assets	measures revenue with respect to investment and is an indicator of asset management
Return on Net Worth	measures revenue with respect to net worth and is an indication of asset management

NOTE: Ratio analysis should always be comparative—that is, individual data should
always be compared to industry data and also to time trend data.

a lower collection period, as well as a lower inventory usage ratio. It
would appear that shelf life is too short for the minority group, which
might be indicative of capital shortage. The ratios are not definitive
enough to warrant a positive statement. The differences in terms of
efficiency and profitability are not significant either way. For the retail
trade area, the choice is negative. The problem here is that MBEs in
this industry are by and large very small firms and, coupled with a high
leveraging factor, this significantly increases the risk of failure.

Wholesale trade. The control group is a somewhat less risky
investment. Financial strength is evidenced by the higher quick and
current ratios and lower financial leverage ratio, although the collection period is longer. Better profitability is evidenced in the higher
return on assets, although the minority group did better in terms of
returns to net worth. This is, however, an indication of lower equity or
assets invested. Note that although the control group does appear

TABLE 4.5 Business Ratios by Industry (Minority)

Ratio Category	Services	Retail	Wholesale	Construction	Manufacturing	Transportation
Quick ratio	1.79	2.64	1.26	1.69	1.89	.40
Current	2.89	8.16	2.76	2.42	3.76	.53
Total liabilities						
Net worth	178.17	178.52	174.91	229.31	178.59	94.27
Collection	32.54	47.04	39.54	73.56	47.22	32.17
period	days	days	days	days	days	days
Inventory						
turnover	113.08	32.88	38.08	68.38	36.97	
Return on assets	21.47	14.43	15.62	15.54	12.32	16.20
Return on net worth	37.29	27.20	31.17	43.82	33.65	41.90

SOURCE: Center for Studies in Business Economics and Human Resources, *Key Business Ratios of Monority-Owned Businesses* (University of Texas—San Antonio), 1981.

TABLE 4.6 Business Ratios by Industry (Control Group)

Ratio Category	Services	Retail	Wholesale	Construction	Manufacturing	Transportation
Quick ratio	2.39	2.05	1.58	2.76	2.16	.89
Current	3.92	6.11	3.49	3.73	3.57	1.47
Total liabilities						
Net worth	151.04	128.58	142.93	155.64	140.49	207.01
Collection	39.41	32.66	48.32	77.93	49.43	31.11
period	days	days	days	days	days	days
Inventory						
turnover	59.93	9.11	26.49	161.32	51.61	287.72
Return on assets	14.51	14.47	16.05	11.64	10.58	5.05
Return on net worth	52.42	28.18	20.34	18.31	10.43	32.80

SOURCE: See Table 4.5.

somewhat more sound, the differences are not that significant so as to preclude the minority group as an investment alternative.

Construction industry. The control group here is definitely more financially sound, more efficient, and more profitable than the minority group from the ratio observations. As an investment alternative, the minority construction industry group does not seem to be a viable alternative.

Manufacturing industry. Here again, the control group compares favorably with the minority group in the areas of financial strength and operating efficiency, although the minority group does perform a little better in terms of profitability. Again, it is not sufficient to only say that the control group appears stronger, but it must be added that the differences are not that significant.

Transportation industry. The control group again appears more favorable than the minority group in terms of financial strength although the minority group is less financially levered. Furthermore, the control group has an unusually high inventory turnover ratio that might be indicative of some problem. The minority group did, however, outperform the control group in terms of the profitability ratio that might have some bearing on investors maximizing expected future benefits from increase in asset valuation.

Because the data are suspect and the differences between the minority and the control group are not that significant, a definitive recommendation cannot be made here. Although in some instances the control group did outperform the minority group, the ratio differentials were not significant enough to preclude minority businesses as investor portfolio choices.

To continue to look at the viability of the minority business sector as an investment medium in conformance with financial risk perspective, an analysis will next be performed on the general failure rate. This is important in that the probability of failure affects investor perspective, especially if the investor wants future income flows.

FAILURE RATES

In 1976, approximately 10,000 firms failed, and in 1977 approximately 8,000 firms failed. Note that at this juncture, failures are only a portion of those firms that ceased to operate. It is important to note that failures refer to businesses that cease to operate and at time of cessation are encumbered with some form of financial liability.

It is difficult to make a fair and accurate comparison of failure rate by ethnic business sectors because of the lack of available data. However, in 1977, the retail industry for the control group experienced a failure rate of 2.4 percent, while manufacturing had a 4.0 percent rate. If this is any indication of general industry behavior, then minority businesses as exemplified in Table 4.5 have a better longevity record than the control group.

Retail industry. Although there are differences between the minority and the control group in the ratio analysis, the failure rate is much lower for the minority group. Depending on investor risk position and portfolio composition, the choice between minority and nonminority firm investment should not be of investor concern.

Manufacturing industry. Because of the insignificant differences between the control and the minority group, the more favorable failure probability of the minority group makes it an investment potential.

For the other groups—wholesale, transportation, and services—the fact that there are no comparable failure data available does not defer from relative lack of variation in those groups in comparison to the control group. The construction areas are vague and the writer will not hazard a stand on that area.

SUMMARY

From an assessment of the criteria for investment and the associated utility functions of investors, a few things are apparent with respect to minority business investment.

(1) Minority businesses are not considered an investment medium and that impact evidences itself in cash scarcity and low equity positions.

(2) Although the ratio data are not statistically sound, by using it as an argumentative base, it is evident that there are no real definite deliniations in financial risk and company life between minority firms and the control group. This therefore implies that investor risk perception is not founded on the basis of business and financial risk, but more on incomplete information and unfamiliarity.

POLICY IMPLICATIONS

In order to develop a larger and more competitive minority owner class, minority youths must be exposed to successful minority business role models at an early age. The ideal situation is some type of "Junior Achievement" or business operation at the high school level. This will make entrepreneurship a viable career alternative.

For those minority individuals already past the first career path choice, accessible information flows on business opportunities, capital sources, and business training are needed. The role model might have been present earlier, but other impediments might have postponed the entrepreneurship choice. This remedy will thus provide the entrepreneurial access.

The fact that minority businesses are excluded from the private capital markets is primarily a result of investor lack of knowledge. Capital market managers must be made aware of this investment area through the promotion of research results and the use of conventional financial institutions and accepted credit methods by existing minority firms. Constant exposure of minority enterprise performance to investment managers will serve to reduce the risk associated with unfamiliarity.

NOTES

1. A numbering system was used to quantify business type so that higher numbers are indicative of low capital costs, high formal education requirements, or a mix of barriers to business entry.

(1) Agriculture & Mining
(2) Construction
(3) Manufacturing
(4) Transportation & Public Utilities
(5) Retail
(6) Wholesale
(7) Finance, Insurance & Real Estate
(8) Service

2. In general, the overall correlation matrix did not justify the role model explanation (Table 4.1), but some of the regional and industry breakdowns did signify the strength of that relationship. For example, Table 4.3, Retail Trade, where the correlation coefficients were significant and positive for the proxy relationships of Urban/Ethnic, TPOP/MPOP, TPOP/MBE, and MPOP/MBE.

3. Recent studies by Peter Bearse, "An Econometric Analysis of Minority Entrepreneurship," Tim Bates, "The Nature of the Growth Dynamic in Emerging Lines of Minority Enterprise," Tim Bates and Antonio Furino, "New Perspectives in Minority Business Development," and David Swinton and John Handy, "The Determinants of Growth of Black-Owned Businesses," show that capital scarcity is still a MBE problem. Start-up capital is limited and very low for MBEs, especially blacks, and also that governmental financial programs have circumvented the private capital market so much so that MBEs are highly leveraged and need more equity financing as opposed to debt.

REFERENCES

BATES, T. (1973) "The potential of Black capitalism." Public Policy 21 (1): 135-148.
———(1974) "Employment potential of inner city Black enterprise." Review of Black Political Economy 4 (10): 59-67.

———(1975) "Trends in government promotion of Black entrepreneurship." Review of Black Political Economy 5 (Summer).

———(1981) "Black entrepreneurship and government programs." Journal of Contemporary Studies (Fall): 59-69.

———(1975) and W. BRADFORD (1975) "An evaluation of alternative strategies for expanding the number of Black-owned businesses." Review of Black Political Economy 5 (Winter): 376-385.

BAUMOL, W.J. (1980) "Toward an operational model of entrepreneurship." Unpublished paper, September.

BEARSE, P.J. (1982) "A study of entrepreneurship by region and SMSA size," in K.H. Vesper (ed.) Frontiers of Entrepreneurship Research 1982. Babson College, MA.

BIRCH, D.L. (1979) "The job generation process." Prepared for the U.S. Department of Commerce/EDA, Cambridge, MA.

BLACKWELL, J.E. (1974) The Black Community: Diversity and Unity. New York: Dodd, Mead.

BOBO, B. and A.E. OSBORNE, Jr. (1976) Emerging Issues in Black Economic Development, Lexington, MA: D.C. Heath.

BOTHA, W. (1981) "Training for entrepreneurs," pp. 153-160 in D.L. Sexton and P. Von Auken (eds.) Entrepreneurship Education. Waco, TX: Baylor University.

BRADFORD, W.D. and A.E. OSBORNE, Jr. (1976) "The entrepreneurship decision and Black economic development." American Economic Review 66 (2): 316-319.

BROCKHAUS, R. (1981) "Psychological characteristics of entrepreneurs," in D.L. Sexton and P.M. Von Auken, (eds.) Entrepreneurship Education 1981 Waco, TX: Baylor University.

BRIMMER, A.F. and H.S. TERRELL (1971) "The economic potential of Black capitalism." Public Policy (Spring): 289-328.

BROOM, H.N. and J.G. LONGENECKER (1966) Small Business Management. Dallas, TX: South Western Publishing.

CAPLOVITZ, D. (1973) The Merchants of Harlem. Beverly Hills, CA: Sage.

Center for Studies in Business, Economics and Human Resources College of Business—University of Texas at San Antonio (n.d.) Minority Cost of Doing Business.

CHEN, G.M. (1979) "Minority business enterprise in the United States: an investment perspective." Houston, TX: South West Finance and Administrative Disciplines Conference, March.

COLES, F. Jr. (1969) An Analysis of Black Entrepreneurship in Seven Urban Areas. Washington, DC: National Business League.

CROMWELL, J. and P. MERRILL (1973) "Minority business performance and the community development corporation." Review of Black Political Economy, 3 (3): 65-80.

CROSS, T.L. (1971) Strategy for Business in the Ghetto. New York: Atheneum.

DINGEE, A.L. and M.A. LEVIN (1980) Formation and Development of Independence Businesses as an Economic Development Tool. Council of State Planning Agencies Working Paper, September.

Dun and Bradstreet, Inc. (1978) The Business Failure Record "1977" Dun and Bradstreet.

Entrepreneurship Institute (1981) New Enterprise and Economic Development Initiatives Today. U.S. Department of Commerce/Small Business Administration/Rockefeller Brothers Fund, September.

FLEXMAN, N. A. (19xx) "Entrepreneurship for career changers." Journal of Career Education 8 (2): 153-160.

GAMBLE, S. (1982) "Minority business survey." Council on Graduate Minority Education/Minority Business Development Agency Student Research Paper, Spring.

HARRISON, B. (1974) Urban Economic Development. Washington, DC: Urban Institute.

HALEY, C. W. and L. D. SCHALL (1973) The Theory of Financial Decisions. New York: McGraw-Hill.

KENT, C. (1981) "The role of entrepreneurship in the private enterprise economy," pp. 1-9 in D. L. Sexton and P. M. Von Auken (eds.) Entrepreneurship Education. Waco, TX: Baylor University.

———D. SEXTON, and S. CONRAD (1981) "Lifetime experiences of entrepreneurs: preliminary analysis," pp. 35-41 in D. L. Sexton and P. M. Von Auken (eds.) Entrepreneurship Education. Waco, TX: Baylor University.

KELLY, P. C. and K. LAWYER (1961) How to Organize a Small Business. Englewood Cliffs. NJ: Prentice - Hall.

KIRCHHOFF, B. A., R. L. STEVENS, and N. I. HURWITZ (1982) "Factors underlying increases in minority entrepreneurship: 1971-1977." Second Conference on Entrepreneurship Research, Babson College, Massachusetts, April.

LEE R. F. (1972) The Setting for Black Business Development. New York: Cornell University.

LIGHT, I. (1972) Ethnic Enterprise in America. Berkeley: University of California Press.

McCLELLAND, D. C. (1960) The Achieving Society. New York: D. Von Nostrand.

———and D. G. WINTER (1969) Motivating Economic Achievement. New York: Free Press.

McFARLANE, C. (1981) "Stimulating entrepreneurship awareness." Journal of Career Education 8 (2): 135-144.

MORSE, R. (1979) "Entrepreneurial initiatives and community need fulfillment," in P.A. Neck (ed.) Small Enterprise Development: Policies and Programmes. Geneva: International Labor Organization.

ONG, P. M. (1981) "Factors influencing the size of the Black business community." Review of Black Political Economy 11 (3): 313-319.

Planco, Inc. (1981) Dimensions of the Hispanic-American Entrepreneurial Experience. Prepared for the Minority Business Development Agency. Dallas, Texas.

RONSTADT, R. (1982) " Does entrepreneurial career path really matter?" in K.E. Vesper (ed.) Frontiers of Entrepreneurship Research 1981. Babson Park, MA: Babson College.

ROSCOE, J. (1973) "Can entrepreneurship be taught?" MBA Magazine June-July.

RUSSELL, J. S. (19xx) "Entrepreneurship: a high priority for business educators." Journal of Career Education 8 (2): 119-126.

SORENSON, A. B. (1975) " Growth in occupational achievement: social mobility or investment in human capital," in K.C. Land and S. Spilerman (eds.) Social Indicator Models. New York: Russell Sage Foundation.

SOWELL, T. (1981) Ethnic America. New York: Basic Books.
SPILLER, E. A. Jr. (1971) Financial Accounting. Homewood, IL: Richard D. Irwin.
WILKEN, P. H. (1979) Entrepreneurship—A Comparative and Historical Study. Norwood, NJ: Ablex.
World Bank (1979) Employment and Development of Small Enterprises. Washington, D.C.

Part II

Federal Economic Development Programs: Analysis and Evaluation

5

Local Discretion:
The CDBG Approach

PAUL R. DOMMEL

☐ DISCRETIONARY DECISION MAKING has been an important part of the history of the community development block grant (CDBG) with economic development funding providing a noteworthy subplot of that history. That subplot illustrates the policy tension that has been the theme of the ten-year history of CDBG—a tension between the process goal of decentralization and the roles of federal and local governments in carrying out the substantive national objectives of the program.

BACKGROUND

CDBG was enacted in 1974, consolidating seven separate discretionary grant programs of the Department of Housing and Urban Development (HUD) into a single formula grant to be distributed to large cities and populous counties.[1] Additional funds were available for small communities through discretionary grants from HUD; in 1981, these small city grants were converted into a block grant to the states. The two largest programs consolidated into the block grant were urban renewal; which had started in 1949, and the Model Cities program enacted in 1966 as a demonstration program for neighborhood revitalization. Since enactment of the program, more than $30 billion in CDBG funds have been distributed.

The focus of this chapter is on the city and urban county entitlement grants. The findings are based on a series of HUD-funded studies conducted over an eight-year period at the Brookings Institution and Cleveland State University (see References).

The principal purpose of the CDBG program, initially proposed in 1971 by President Nixon as special revenue sharing, was to shift a greater share of decision-making authority from federal to local officials and to provide the latter with more flexibility in the use of the funds. But besides this process-oriented goal, the legislation as enacted in 1974 also included several substantive goals, the two most important ones being eliminating or preventing slums and blight, and assuring that the primary benefits of the program went to low- and moderate-income groups (Public Law 93-393, 1974).

The block grant program was extended three times for three-year periods. In 1977, the extension also included a major change in the funding system to provide a larger share of the money to the older, declining cities of the Northeast and Midwest (Public Law 95-128, 1977). In 1980, the program was again extended for three-years with no major changes made in the substance of the law (Public Law 96-399, 1980). A year later, in 1981, as part of President Reagan's program, the block grant underwent a major procedural revision with the intended effect of greatly reducing federal involvement in CDBG decision making, which had increased appreciably under President Carter.[2] The most recent extension came in 1983 (Public Law 98-181).

These legislative milestones, along with their implementation by three different administrations, had important effects on the use of CDBG funds for economic development activities.

ECONOMIC DEVELOPMENT—OPPORTUNITIES AND CONSTRAINTS

As enacted in 1974, economic development was not established as a significant area of anticipated development activity for block grant funding. The law provided a long list of activities eligible for assistance, including all activities that had been authorized under any of the categorical grants that had been consolidated into CDBG. Within this listing were activities that could be interpreted as supporting economic development, but economic development was not set forth as a distinct program category.

This was changed in the 1977 amendments that added a paragraph to the list of eligible activities to include a range of economic development activities.[3] However, such assistance could not be direct but rather had to be provided through neighborhood-based organizations, a local development corporation, or a small business investment company. Such organizations had considerable flexibility in the kinds of assistance they could provide and they could undertake activities that the recipient city or county itself could not directly fund. Thus, economic development under CDBG was operated through intermediaries, some of which were dependent upon CDBG as the primary source of operating and investment funds.

While expanding the visibility and flexibility of the block grant to support economic development activity, the 1977 amendments included a more important change affecting economic development. The legislation established a new discretionary grant program, the Urban Development Action Grant (UDAG), which initially authorized $400 million a year to undertake development activities in severely distressed cities and urban counties. The federal assistance was to be used for leveraging private sector funds for a variety of development activities, including economic development.

The 1981 amendments of CDBG went further and permitted recipient communities to make the block grant assistance available directly to for-profit organizations for economic development.[4]

Thus, since 1974, economic development evolved *legislatively* into an integral part of the block grant program, providing local participants in the decision process more opportunities to fund such activities.

The word "legislatively" was emphasized since that expanding statutory basis for local flexibility on use of the block grant was not always matched by the policy preferences of federal program administrators. Nor did the added flexibility provided for economic development activities necessarily mean a corresponding growth in support of such allocations at the local level.

FEDERAL POLICY PREFERENCES

Viewing the decade since the program began, the general evolution of federal policy toward CDBG has been characterized as a shift from federal "hands off" during the first two years under President Ford, to "hands on" during the four years of President Carter, and back to a more determined "hands off" policy under President Reagan (Dommel et al., 1982: Chs. 2, 9).

The hands-off approach of the Ford administration reflected a policy preference for the process goal of decentralization, thereby minimizing federal oversight of the program and allowing local participants maximum flexibility in establishing development strategies and spending priorities. Within this policy context, the general programmatic result initially was considerable continuity in development choices for those communities that had participated in the previous categorical programs; that is, a large share of funds went to continue work on urban renewal projects in the downtown area or in residential neighborhoods, and to continue funding a variety of social services started under the Model Cities program. Also, the added flexibility enabled the communities to expand revitalization activities into new neighborhoods that had not previously received benefits from the defunct categorical grants. For some communities that had not participated previously in the categorical programs, the block grant provided a new source of funds to undertake revitalization programs that were generally modeled on the previous federal approaches. For both the experienced communities and those receiving HUD funds for the first time, the prior federal categorical programs provided a basic model that local participants tailored to meet diverse local development needs and political demands.

In the early period of CDBG, decision making and establishing development priorities, economic development generally did not fare well. While some individual communities gave high priority to economic development, the overall commitment of CDBG funds to such activities was quite low. A study by the Brookings Institution of 60 jurisdictions across the country showed that during the first two years of CDBG, economic development was the smallest of six program categories funded, both in terms of the number of sample communities making such allocations and the amount of money committed. Overall, only 3 to 4 percent of the total dollars received by the sample communities was earmarked for economic development, although a few of the sample cities such as Chicago, Rochester, New York, and East Orange, New Jersey allocated 10 percent or more for this purpose (Dommel et al., 1978: 189-192). The kinds of activities funded ranged from technical assistance for small businesses to development of an industrial park.

The low level of economic development allocations in the early years of the block grant appeared to result from the need to continue funding earlier program commitments with established political and bureaucratic support and/or the need to respond to new demands for more politically popular and visible activities such as housing rehabilitation and neighborhood public works. Economic development activities, which often had longer-term and frequently only potential prospects of yielding tangible benefits, could not compete effectively. Also, the uncertainty of what economic development activities could be funded under the law put economic development at a competitive disadvantage.

The statutory changes made in 1977 provided a more firm foundation for those seeking to give higher priority to economic development. Also, the formula change in that year provided a number of cities and counties in the Northeast and Midwest with a major increase in funding, providing a new source of uncommitted funds that might be channeled to economic development activities. This in fact did occur in some cities. The Brookings study stated that several cities in its sample earmarked part of their additional money for economic development activities. The overall finding of that research was that, as the program matured, the number of communities making allocations for economic development increased, as did the share of funds going for such activities (Dommel et al., 1980: 137-140). However, the overall share going for economic development continued to be relatively small and remained at the bottom of the program priorities, receiving about 7 percent of the sample funds in the fourth year of the program.

Local preferences for revitalization of residential neighborhoods and for near-term results remained the primary determinants of this low priority, but federal administrative policy preferences also had a dampening effect on any potential local interest in economic development funding through the block grant.

Soon after taking office, officials of the Carter administration adopted a more activist policy of federal involvement in CDBG, shifting the policy preference from procedural decentralization to greater federal oversight of the substantive national objectives.

Among the national objectives, HUD officials chose to give particular emphasis to the provisions of the law to direct the primary benefits of the program to low- and moderate-income groups (Embry, 1977; Federal Register, 1978: 8434-8447). Accompanying this policy of targeting benefits was a policy of geographic targeting; that is, the concentration of activities in residential neighborhoods so that visible improvements could be achieved within a forseeable period of time.

The combination of these two policy choices made it more difficult to qualify economic development activities for funding. A specific constraint was the policy test for the distribution of benefits from economic development projects. The new HUD policies implemented by the Carter administration provided that economic development projects funded through CDBG had to result in jobs that would go to low- and moderate-income persons. HUD closely scrutinized local economic development proposals for compliance with that policy.

Partly as a consequence of this federal policy and the lack of local support for such activities, CDBG funding of economic development remained relatively low, reaching just 9 percent of total sample funds in the sixth year of the program, according to the Brookings study. However, this relatively small absolute growth represented the greatest rate of increase (388 percent over six years) in allocations among all program categories (although its rate of increase was largely a function of the small amount of the initial allocations). The increases, made possible in part by the 1977 legislative and formula changes, moved economic development from the smallest category of dollar allocations in the early years to fourth place in year 6. Also, the number of communities making such allocations increased from 21 to 36 of the 60 sample communities over the six years (Dommel et al., 1982: 73). This increase in priority for economic development was accompanied by a drop in allocations for continuing urban renewal projects started under the categorical programs; such projects, some of which had economic development objectives, were either completed or stretched out.

It is worth noting that by the sixth year of the program, several cities in the sample were now making substantial allocations for economic development. In year 6, six cities and one urban county in the sample earmarked at least 20 percent of their funds for such activities (Dommel et al., 1982: 123-134). The kinds of projects funded continued to range over a wide variety of activities, now including rehabilitation of neighborhood commercial areas. This particular use is worth noting since it indicated that activities classified as economic development did not necessarily mean the creation of new jobs or even saving jobs. Revitalization of neighborhood commercial areas fit the general neighborhood orientation of HUD policy during the Carter years and as such was an eligible economic development activity. Job-creating economic development activities offered more potential through the UDAG program also administered by HUD.

The coming of the Reagan administration and a change in the national economic environment within which CDBG operated created new opportunities for using CDBG for economic development projects.

POLICY DEREGULATION AND ECONOMIC DEVELOPMENT

The Reagan administration came to office advocating major deregulation of a wide range of domestic policies, including CDBG. Such deregulation was to be both administrative and legislative (Dommel et al., 1983: Ch. 2). For CDBG, the administrative deregulation was to be implemented through fewer and more flexible regulations and guidelines. Legislatively, the federal role was to be reduced primarily through elimination in the law of the formal CDBG application process, whereby HUD had the opportunity to say no to entire locally proposed programs or to individual activities.

Reagan's deregulatory approach represented a major change in the policies of the preceding administration that had imposed greater federal regulatory oversight to achieve its benefits and geographic targeting objectives. The new policies signaled the HUD area offices and community participants that the rules of the CDBG game were changing in important ways.

The stated general intention of the new administration, which took office in 1981 just as most communities were submitting their applications for year 7 funds, was to "remove some review requirements, simplify the application review process, and reduce costs of government" (HUD Notice, 1981: 1). The primary beneficiaries of the deregulatory policy were to be local officials who would have greater flexibility in allocating CDBG funds that, however, were to be cut as part of the president's overall fiscal plan to reduce the budgetary costs of a wide range of domestic programs.

Three important features of the new flexibility were (1) removal of any percentage measure of the share of benefits that were to go to lower income groups; (2) wider latitude in where CDBG funds could be spent in a community by dropping the geographic targeting policy; and (3) allowing communities to provide CDBG-funded public services outside of designated target areas.

The guidelines also increased local flexibility for funding economic development activities. As noted earlier, the regulatory policy under the Carter administration had been that the basis for eligibility of economic development activities was whether or not the jobs created or saved would go to lower-income persons. The 1981 guidelines stated that "this type of judgment is often impractical and has proven unduly restrictive" (HUD Notice, 1981: 4). The new criterion for eligibility was whether the jobs would *be available* to such persons, not that they necessarily would go to persons in those income groups. These guidelines gave local officials a sense of direction for the new deregulatory policy preference. It was not long before these administrative guidelines were given added substance by legislative changes proposed by the Reagan administration and adopted by Congress.

The single most important change in CDBG made by Congress in the summer of 1981 was elimination of the application procedure that the Carter administration had used to influence local program strategies at the planning stage of the CDBG process. By eliminating the application, the Reagan administration sought to shift the burden of local program accountability from the planning to the performance or implementation stage of the CDBG process. That is, HUD's principal involvement in local programs was statutorily shifted from a prospective look at local development strategies through the application process to a retrospective assessment of individual activities and implementation performance. This, combined with the change in the guidelines, meant that significant new opportunities had opened up for using the block grant for economic development.

Some early findings on the effects of the deregulatory policies on CDBG allocations indicated that economic development activities gained a higher priority at the local level. A study by Cleveland State University of ten large entitlement cities found that these cities, which had been allocating an average of 8 to 9 percent of their funds for economic development throughout most of the program, increased that average to 12 percent in year 8 of the block grant, the first year when both the administrative and legislative deregulation were in effect (Dommel et al., 1983: 75). Seven of the ten cities increased their funds for such activities. A point to be noted is that in year 8, the total dollar allocations for economic development in the ten cities increased by 28 percent while total program funds declined by 12 percent primarily because of the federal budget cuts. However, it must also be noted that the dollar increases in the sample were largely accounted for by large increases in three of the ten cities and that, for the sample as a whole, economic development was only the fourth largest program category in terms of total dollars allocated.

An analysis by HUD of a sample of 200 entitlement communities reported a similar upward trend in economic development allocations. HUD reported that funding increased from 5 to 8 percent of total CDBG funds between years 7 and 8 (HUD Annual Report, 1983: 23). The larger percentage of the Cleveland State study indicated that the larger cities in that sample tended to allocate larger shares of their funds for economic development.

Thus, the early data after the new policies went into effect suggested that cities were giving increased priority to economic development activities. This appeared to be linked to the greater flexibility afforded by the deregulatory policies as well as the general state of the national economy with economic development becoming a new "motherhood" issue.

At the local, new mayors in several cities in the Cleveland State study were promoting economic development as a top priority and turned to the CDBG program as a vehicle for carrying out their commitments (Dommel et al., 1983: 85-87). One of the most evident cases was in Atlanta where a new mayor took office soon after the deregulatory policies had been applied and moved to establish new priorities for the city's CDBG program. While Atlanta was one of those cities that had been allocating above-average amounts of the block grant for economic development in the past, the city's top

CDBG priority consistently had been neighborhood revitalization focused on housing rehabilitation and neighborhood public works. The new city administration set a 35 percent funding goal for economic development and allocations increased from 11 percent in year 6 to 26 percent in year 8 as the mayor's priorities began to be implemented.

In Houston, another new mayor also put a high priority on economic development. There, $2.3 million, 10 percent of the city's grant, was earmarked for economic development in year 8. This was the first time since 1976 that city officials had included funds for economic development in the CDBG program. The city intended to use its CDBG funds to provide public improvements in the city's newly created tax-increment finance district near downtown Houston and to supplement the city's industrial revenue bond program that had been enacted recently to create jobs and bolster the tax base in the city's CDBG target areas.

It is worth noting that where economic development had been a city priority or where it more recently gained a significant priority, the chief source of support tended to be from the city's executive branch, primarily the mayor. This was most evident where the election of a new mayor led to a new priority for CDBG-funded economic development activity. The indication is that CDBG-funded economic development activities, which often lack political and citizen support from the bottom, are dependent on sponsorship from the top.

LEVERAGING WITH CDBG FUNDS

It is also reasonable to speculate that one of the reasons for the general increase in economic development funding under the block grant is that it is one of the few programmatic categories that can be used to leverage private sector funding, thereby increasing the total impact of the CDBG program. The other major program category where this is possible is housing rehabilitation where additional non-public funds can be obtained from private lending institutions. As CDBG funds continue to decline in nominal and/or real dollars, use of block grant funds to leverage private sector dollars takes on added importance and such activities may get increased priority or at least be maintained at current funding levels.

In addition to providing direct assistance for economic development, there are other provisions of the CDBG law that have enabled some cities to use their block grant money to leverage additional resources for economic development (Dommel et al., 1983: 88-89).

Under Section 108 of the law, a city may borrow up to three times the amount of its entitlement from the federal government for eligible short-term community and economic development activities. The most frequent use of the money has been for land acquisition and related site preparation costs. Some cities have used this provision to provide short-term loans to the private sector for economic development activities and have pledged their future CDBG entitlements to guarantee these loans. If the private sector defaults on the loans, the funds are repaid to HUD from the city's future CDBG entitlement.

The Cleveland State study reported that half of the ten cities (Atlanta, Cleveland, Phoenix, Rochester, and St. Louis) in that sample had initiated Section 108 projects. Atlanta used $3.5 million in Section 108 funds for public improvements for a 178-acre industrial park currently being developed in northwest Atlanta. Other funds to be used in this project included direct CDBG grant assistance and a $1 million grant from the U.S. Economic Development Administration. In Cleveland, Section 108 money is being used to create a revolving loan fund to aid the private sector in revitalizing the "flats area," the city's former warehouse district that is now undergoing commercial and residential revitalization. The Section 108 funds will be loaned to private developers at below market rates to finance the acquisition and renovation of vacant warehouse structures. In Phoenix, a $10.8 million Section 108 loan was executed in year 7 to be used over a three-year period for a major residential/office development in downtown Phoenix. Of that amount, $6.6 million was earmarked for land acquisition and subsequent development (e.g., relocation, demolition, clearance, public improvements) prior to its disposition to a private developer, an approach similar to the former urban renewal program. Phoenix officials estimate that the amount of private funds leveraged from this project will be in excess of $130 million.

Another CDBG leveraging device is the "CD float," which allows cities to draw down their unexpended funds from previous CDBG grants to create a capital pool for short-term use by the private sector in economic development projects. The funds have generally been used for interim financing during the construction phase of development projects. The loan paybacks are then used for the CDBG projects for which the funds had been allocated originally and the city uses the interest earned for additional community development activities eligible under the CDBG law. Among the cities in the Cleveland State sample, Rochester and Seattle have initiated CD float projects.

The attractiveness of both the Section 108 and the CD float is that both techniques allow CDBG entitlement cities to leverage private investment for additional community development activities and provide a means for cities to generate additional revenues during a period of fiscal austerity.

In year 9, Seattle drew down its entire letter of credit ($12.5 million), which was then loaned at a below-market rate to a private developer for use in a major downtown redevelopment project. In return, the developer agreed to provide about $300,000 in public works and also pay nearly $1 million in interest on the city loan. City officials noted that because of the interest income from their float loan they were able to lessen the impact of their reduced entitlement grant.

Rochester used the proceeds from two Section 108 loans to establish a business loan program under which the city lends up to a maximum of $500,000 to private firms to assist in financing expansion and renovation. In 1981 and 1982, the city loaned about $3.2 million to private firms through this program.

Rochester has also executed three float loan projects. The first was a $2 million low-interest loan for a major downtown office building. A second float loan of about $1 million was executed with a developer of a downtown hotel. A third loan of $300,000 was designated for the rehabilitation of an office building that in turn is to serve as a catalyst to encourage other developers to renovate additional buildings in the project area.

CONCLUSION

Economic development under CDBG has evolved into something of a growth industry since the program first started in 1974, but its growth potential overall is likely to be a limited one. The data indicate that while some economic development (broadly defined) aid is characteristic of many local programs, relatively few entitlement jurisdictions allocate a significant share of their grants for this purpose. This is not surprising since local participants in CDBG decision making, officials and citizens, have preferred to channel the largest share of funds into the revitalization of residential neighborhoods, public services, and general public improvements that might normally be funded from local tax revenues. The recent funding constraints in CDBG make it more difficult for advocates of economic development to compete effectively with established priorities.

Another factor likely to limit the growth of block grant funds for economic development is the distribution of benefits that has been a continuing and important issue of the program since its beginning. Although the substantive national objective of directing the primary benefits of the program to lower-income groups is no longer given its former emphasis by HUD, neither federal nor local officials are likely to risk the political consequences of significantly downgrading that legislative policy goal. Within that context, since economic development activities tend to have a lower level of such benefits than many of the other eligible activities of CDBG, the funding of economic development is likely to be dealt with cautiously by local officials. The greatest potential for economic development appears to be in those communities where key elected officials choose to press for such development generally and where CDBG can be used as part of that effort, particularly to leverage private sector funds.

The relatively low priority given to economic development under CDBG should not be interpreted negatively. The block grant was not created as an economic development program, as reflected by the legislative history of CDBG and by the creation of UDAG to serve such a purpose more directly. Nevertheless, the block grant is sufficiently flexible and, through the Section 108 and CD float devices, supportive of economic development to serve as a useful tool for interfacing with other elements of a local economic development strategy.

NOTES

1. For the general policy history of CDBG, see Nathan (1977): Ch. 2; and Dommel et al. (1982): Chs. 2 and 9.
2. These CDBG amendments were a part of the "Omnibus Budget Reconciliation Act of 1981."
3. See Public Law 95-128 (1977), Sec. 105 (a) (14).
4. See Public Law 97-35 (1981), Sec. 105 (a) (17).

REFERENCES

DOMMEL, P. R., R. P. NATHAN, S. F. LIEBSCHUTZ, M. T. WRIGHTSON, et al. (1978) Decentralizing Community Development. Washington, DC: U.S. Department of Housing and Urban Development.
DOMMEL, P. R., V. E. BACH, S. F. LIEBSCHUTZ, L. S. RUBINOWITZ, et al. (1980) Targeting Community Development. Washington, DC: U.S. Department of Housing and Urban Development.
DOMMEL, P. R., J. MUSSELWHITE, S. F. LIEBSCHUTZ, et al. (1982) Implementing Community Development. Washington, DC: U.S. Department of Housing and Urban Development.
DOMMEL, P. R. et al. (1982) Decentralizing Urban Policy: Case Studies in Community Development. Washington, DC: Brookings Institution.
DOMMEL, P. R., M.J. RICH, L.S. RUBINOWITZ, et al. (1983) Deregulating Community Development. Washington, DC: U.S. Department of Housing and Urban Development.

NATHAN, R. P., P. R. DOMMEL, S. F. LIEBSCHUTZ, M. D. MORRIS, et al. (1977) Block Grants for Community Development. Washington, DC: U.S. Department of Housing and Urban Development.

EMBRY, R. C., Jr. (1977) Memorandum to CDBG field staff, "Management of the Community Development Block Grant Program." April 15.

Federal Register (1978) Vol. 43, March 1.

Public Law 93-383 (1974) Housing and Community Development Act of 1974.

Public Law 95-128 (1977) Housing and Community Development Act of 1977.

Public Law 96-399 (1980) Housing and Community Development Act of 1980.

Public Law 97-35 (1981) Housing and Community Development Amendments of 1981.

U.S. Department of Housing and Urban Development (1981) Notice CPD 81-5, Community Planning and Development, "Review of Community Development Block Grant (CDBG) Entitlement Applications." May 15. Washington, DC: Author.

———(1983) Consolidated Annual Report of Congress on Community Development Programs. Washington, DC: Author.

6

Competitive Grants:
The UDAG Approach

PAUL K. GATONS and MICHAEL BRINTNALL

□ THE URBAN DEVELOPMENT Action Grant (UDAG) program was created in 1977 to complement and expand the revitalization activities of the Community Development Block Grant (CDBG) program.[1] The program was to target federal resources to communities with greatest need for economic and neighborhood revitalization by stimulating private investment in specific commercial, industrial, and neighborhood development projects. In addition, it was intended to overcome problems of previous economic development programs through heavy private sector involvement in "public-private partnerships."

UDAG is built around the concept that a project faces a definable gap in financing or development (e.g., the cost of money is too high or a sewer hook-up is too far away), and that the public and private sector will reach a contractual agreement, on a case-by-case basis, about a subsidy to close the gap. This contrasts, for example, with the Enterprise Zone concept that starts with an assumption that a development gap in distressed areas has many dimensions (e.g., high development and operating costs, regulatory burdens, low public services, crime, and a discouraging quality of life) that must all be addressed simultaneously so that one problem does not overwhelm remedial efforts for another; and that a multi-faceted range of remedies such as tax breaks and improved public services should be provided as incentive available to anyone willing to locate in the distressed area.[2]

AUTHORS' NOTE: *The statements and conclusions in this chapter are solely those of the authors and do not necessarily reflect the views of the U.S. government or the Department of Housing and Urban Development.*

The general application of the UDAG program has been shaped by four pieces of legislation: the Housing and Community Development Amendments of 1977, which provided the framework for the program; the Community Development Amendments of 1979, which expanded eligibility to nondistressed cities through a "pocket of poverty" provision; the Omnibus Reconciliation Act of 1981, which placed greater emphasis on economic recovery and job creation; and the Housing and Urban-Rural Recovery Act of 1983, which reauthorized the program and included provisions that were designed to avoid the exclusion of housing projects and insure the continuation of the program as a centralized competition.

The program's success and popularity is attested to by its survival in the face of a strong movement to entitlement block grants and decentralized decision making and responsibility. It has evolved into a program with substantial flexibility to meet a variety of local development preferences and strategies, and to fund all types of projects, with tailored subsidies, in communities of all sizes.

OVERVIEW OF PROGRAM OPERATION

As discussed later in more detail, a community may apply for a grant if it meets either program standards for physical and economic distress or has a "pocket of poverty." It must also have demonstrated results relating to low- and moderate-income housing and equal opportunity in housing and employment. For funding considerations, communities are separated into "large" and "small" categories, with the former consisting of all central cities regardless of population size, cities with populations over 50,000, and urban counties, and the latter, for which a 25 percent set-aside of funds is provided, composed of all noncentral cities with populations under 50,000.

Applications submitted by eligible communities are reviewed by staff in HUD's field and headquarters offices. Besides meeting minimum program requirements for firm private and public commitments and the ratio of private investment to UDAG funds (2.5 to 1), to be considered "technically fundable," projects must meet the "but for" requirement ("but for" UDAG, the project could not be completed), request the least amount necessary to complete the project, and be "feasible." Substantial negotiations between HUD, the applicant, and the private sector participants may occur before a project is determined to be technically fundable. Technically fundable projects are forwarded to the secretary for final funding decisions.

A decision to fund a project constitutes only a preliminary award. For a project to receive the funds, legally binding private and public commitments must be provided and a legal grant agreement that spells out all action and performance requirements must be executed by the grantee and HUD. Once underway, projects are monitored by HUD field staff and progress reports from the grantees are required quarterly. A project is "closed out" when all activities defined in the grant agreement are finished and all costs incurred. A project is not subsequently considered "completed" until all performance requirements are essentially met and a final audit has been approved.

TARGETING TO DISTRESSED COMMUNITIES

From the beginning of the program through the end of FY 1983, 929 communities have been awarded over $3 billion in UDAG funds for more than 1,900 projects.[3] About 80 percent (294) of all eligible large cities and urban counties have received awards. New York City and Baltimore each have been granted more than 50 UDAG awards, while a number of other cities—Chicago, Cleveland, Detroit, Rochester, New York, and Boston—each have more than 20 projects. Over 600 small cities have UDAG projects at various stages of completion. Awards range in size from $40 million to Jersey City, New Jersey for Phase I of a major mixed-use development (retail center and 1,000 rental housing units), down to $33,200 to help finance the construction of four townhouses in LaGrange, Georgia.

On a statewide basis, communities in New York are by far the largest users of the program with 291 awards for over $330 million. Communities in three other states—Michigan, Pennsylvania, and Massachusetts—each have obtained more than $200 million in grants. In contrast, there is only one project in Idaho with a $700 thousand UDAG. As to the distribution among census regions, not surprisingly with the heavy program users mentioned above, the Northeast and North Central regions have received the most UDAG funds. About one-third of the UDAG dollars has flowed to each of these regions, while somewhat more than one-fifth of the dollars has been received by communities in the South and slightly less than 10 percent by those in the West. Of course, the number, size, and degree of distress of eligible communities differ between regions. The regional distribution of population in eligible communities provides a rough indicator with which to compare the regional distribution of funds. On this basis, the population in eligible communities provides a rough indicator with Northeast and North Central regions, each with about 28 percent of the eligible population, have received a slightly higher amount of funding in proportion to their populations, while the South and West, with 30 percent and 14 percent of the eligible population, respectively, have received slightly less.

Perhaps of greater importance than the distribution of awards among large and small cities, states, and regions is the extent of targeting to the more distressed or "needier" communities. One method used by HUD to rank eligible communities by degree of distress is the "impaction index," applied separately to large and small cities. An impaction rank is determined for each community by a weighted index of three of the eligibility criteria: age of housing, poverty, and population growth/lag. For purposes of examining the extent of targeting, it is useful to categorize grantees into three equal groups as highly, moderately, or less distressed. Applying this method to applicants for all projects receiving awards through FY 1983, it can be concluded that UDAG awards have been heavily targeted to the one-third highly distressed eligible large cities and urban counties. About 62 percent of both projects and dollars are attributed to this group, while 25 percent of awards and dollars have gone to the moderately distressed one-third, and only 13 percent to the less distressed one-third. However, the targeting picture is different for small cities. The highly distressed one-third of small cities received less than one-third of the small cities project awards and 34 percent of the small cities UDAG dollars. Those in the moderately distressed category accounted for 27 percent of the projects and dollars, while small cities classified as less distressed accounted for about 40 percent of the projects and 39 percent of the funds.

The program's ability to target assistance to communities with the greatest need depends on how eligibility is determined, the selection process, and whether the neediest cities submit project applications.

ELIGIBILITY

The degree of targeting can be varied in an eligibility system depending on the specific criteria chosen to reflect distress, the minimum standard set for each criterion, and the number of minimum standards that must be met or exceeded. Six criteria were chosen to determine program eligibility: percentage of housing constructed prior to 1940, per capita income change, percentage of poverty, population growth lag/decline, job lag/decline, and unemployment rate. These six criteria were selected because they were either direct measures or proxies for the eligibility factors listed in the 1977 Act: age and condition of housing including residential abandonment, average income; population outmigration, and stagnating or declining tax base.[4] A seventh criterion (discussed later)—location in a labor surplus area (LSA)—will be added to the program in 1984.

The minimum standard or threshold for each criterion is the median value for all large cities. Thus, half of the large cities will receive credit (an eligibility point) for meeting or exceeding each standard. In general, communities must receive at least three eligibility points to be designated as "distressed." Large cities and urban counties must meet three of the six (or seven, when LSAs are added) minimum standards; if the poverty rate is less than half the median value, four of the standards must be met (in effect, a penalty point is levied for low poverty). If the eligibility requirement were raised to four points, eligibility would decline from about half of all large cities to less than 40 percent.

Unemployment data are not available for small cities and job lag/decline data are not provided for cities of less than 25,000. Small cities above 25,000 population must meet three of the available five standards; as with large cities, a penalty point may be applied for a low poverty rate, but, in addition, a "bonus point" may be awarded. If the percentage of poverty is twice the standard the community must meet only one additional standard. If the percentage of pre-1940 housing is twice the standard, the community need only meet the poverty standard. Cities of less than 25,000 population must meet three of the available four standards or may receive, as with the larger small cities, bonus points for high poverty or age of housing.

The medians for large cities are used as the minimum standard or threshold for small cities. This system probably benefits small cities, at least on some factors. For example, with populations relatively stable at least until the 1970s, small cities can be expected to have a high percentage of housing stock built before 1940 (the average for all nonmetropolitan urban places has been about one-third again higher than in metropolitan cities). Likewise, although the average incidence of poverty is higher in small cities, poverty is a national measure and doesn't consider differences in the cost of living between large and small cities. Finally, the distress standard for change in per capita income is measured in absolute dollar increases. Given the smaller income base found in many small cities, income gains that are small enough to meet the large city based standard may represent relatively high percentage gains. In general, minimum standards based on small city median values would likely reduce the number of small cities qualifying as distressed.

Cities that do not meet program distress criteria may still apply for UDAG awards through the "pockets of poverty" provision. Essentially, pockets of poverty are areas within a city with concentrations of low-income persons and those in poverty. There are minimum population requirements to qualify as a pocket and the city must certify that it is providing basic services in the area that are at least equivalent to services provided in more affluent areas. Local matching funds equal to 20 percent of the grant requested must be provided for the project, and there are requirements relating to direct benefits to low- and moderate-income residents of the pocket, particularly employment opportunities. By the end of FY 1983, 29 awards had been made under the pockets provision.

Indian tribes are also considered in the eligibility system. A tribe is presumed to meet the minimum standards of distress unless data made available indicate that the tribe's distress is not comparable to that of eligible small cities. So far, two UDAGs have been awarded to tribes—the Navapoi-Prescott Tribe of Arizona to construct a hotel on the reservation and the Mississippi Band of Choctaw Indians for expansion of a manufacturing facility.

Finally, even if a community meets all of the eligibility distress criteria, to become fully eligible to apply for a grant, it must demonstrate results in providing housing for low- and moderate-income persons and equal opportunity in housing and employment. Hundreds of small cities have been denied full eligibility on this basis, as well as large cities such as Philadelphia and San Francisco in the early years of the program.

LABOR SURPLUS AREAS

Two factors have led to the only major change in eligibility determination for distressed communities since the beginning of the program—use of labor surplus areas. First, the UDAG amendments in the Omnibus Reconciliation Act of 1981 have been interpreted by the HUD administration as presenting a strong shift in emphasis to job creation. Rather than the 1977 Act's purpose of "alleviating physical and economic deterioration in severely distressed cities and urban counties" (which was widely accepted as meaning establishing viable communities through economic revitalization), the 1981 Act established that the grants were to be made "to cities and urban counties which are experiencing severe economic distress to help stimulate economic development activity needed to aid in economic recovery."

The second factor that led to the eligibility change was the introduction of data from the 1980 census in determining minimum standards for eligibility. Updating data from the 1970 census, particularly the poverty measure, contributed to dropping a substantial number of communities from eligibility (over 2,000 small cities), many of which

were experiencing high unemployment rates (which was not considered in small city eligibility determination). Many small places lost
the eligibility points for poverty when 1980 census data were introduced because poverty rates had previously been calculated as a ratio
of the 1970 number of poor persons to annual estimates of total
population. This technique resulted in a poverty rate that was decep-
tively high for communities losing population, since the denominator
was adjusted downward annually, but not the numerator. More accurate data in 1980 caused the calculated poverty rate to drop.

A number of options to address the greater focus on economic
development and the high unemployment rates of the cities facing
ineligibility were discussed internally. These options ranged from
double weighting unemployment to the elimination of all eligibility
criteria except unemployment (unemployment was originally chosen
as a proxy for a stagnating or declining tax base). Of course, a major
problem with most of the unemployment options considered was the
absence of such data for small cities. Thus, the solution arrived at was
the introduction of an additional criterion of economic distress: location in a labor surplus areas (LSA).

LSAs are large cities, counties, and county balances with high
unemployment, designated by the Labor Department for the purpose
of targeting federal procurement. An area is designated an LSA if it
has more than 120 percent of the national average unemployment over
the last two years. The 120 percent threshold is limited so as not to
exceed 10 percent or go below 6 percent. Large cities will now be able
to attain eligibility by meeting three out of seven (rather than six)
criteria. Thirteen more large cities will become eligible with this
change. Small cities will use county or county balance LSA data as a
measure of their unemployment problem. About 3,000 more small
cities will become eligible with this change, more than making up for
the number losing eligibility from the application of 1980 census data.

PROJECT SELECTION

Given the large number of eligible communities, the selection
process becomes an important determinant of the extent to which the
program is targeted to the more distressed communities. To be considered for project selection, there must be a firm private financial
commitment and a firm commitment of public resources (if included in
the project). Firm commitments are generally letters for project financing that commit the participating parties to specific activities and
must show that investment will occur contingent only upon the award
of the UDAG. In addition, projects are required to have a minimum

leverage ratio (private investment to UDAG dollars) of 2.5 to 1.0. In selecting among projects, both community and project factors are to be considered. The primary criterion for selection is to be the comparative degree of economic distress among applicants. Other selection factors to be used in choosing among applications relate to employment impacts, particularly on lower-income persons and minorities; the leverage ratio; fiscal impacts; impact on physical conditions of the community; likelihood of completing the project in a timely fashion; demonstrated performance in carrying out housing and community development programs; extent of relocation; and so forth.[5] Finally, the project must meet the "but for" condition ("but for" the UDAG, the project would not be undertaken) and not substitute UDAG funds for local funds, and the grant amount must be the least amount necessary.

Applications are accepted for separate large and small city funding rounds each quarter. The applications are assigned to program staff who perform an underwriting function to determine whether the project is "technically fundable" or can be made technically fundable. In doing so, they concentrate on four factors: the "but for" requirement; whether the amount requested is the least necessary to complete the project; the leverage ratio; and project feasibility. In addition, the selection factors mentioned above may affect fundability (e.g., an industrial project that creates no new permanent jobs). Extensive and intensive negotiations may be conducted with the applicant and the private sector participants, often resulting in changes to the proposal. Applications that may be made fundable but that require new commitments because of project changes may be held over to later rounds. Only applications considered technically fundable are forwarded up the decision line. The UDAG Office Director and the Assistant Secretary for Community Planning and Development must also pass judgment on a project's fundability before it is forwarded for the secretary's consideration. In many cases, the community officially withdraws its application, thus avoiding the "rejection" label. Projects not technically fundable are placed in the "no further consideration" (NFC) category. Besides withdrawal, the most often indicated reasons for NFC are insufficient financial commitment, failure to meet the "but for" test, and incomplete application.

FUNDING SUCCESS RATES

The funding success rate (percentage of applications receiving awards) for all UDAG applications submitted to HUD has been 50 percent. Large cities have been relatively more successful than small cities in getting applications funded (56 percent versus 44 percent success rates). Success rates for large cities and urban counties relative to the degree of a community's distress show surprisingly small variation (58 percent for the highly distressed group, 55 percent for the moderately distressed, and 51 percent for the less distressed). The heavy relative funding (targeting) to projects in the highly distressed large cities indicated earlier results more from the volume of applications received from these communities than from funding success rates. Indeed, these communities' share of awards closely parallels the share of applications submitted, for example, highly distressed communities submitted 61 percent of the applications and received 62 percent of the awards.

Success rates for small cities relative to the degree of distress also show small variation, but in contrast to large cities, the less distressed group had a somewhat higher success rate (highly distressed—42 percent; moderately distressed—43 percent; and less distressed—50 percent). Application submission was more evenly distributed among the three small city distress groups than was true for large cities, which in turn was again reflected in the more even distribution of awards among the groups.

Although half of the applications submitted received funding, the success rate of applications that were determined to be technically fundable has probably been closer to 100 percent. In most years of the program and in most funding rounds, the demand for funds from fundable projects has not exceeded the available supply of UDAG dollars, particularly in the case of small city projects. In the past, when demand for funds did exceed the supply, awards in a round generally went to the cities with the highest impaction rankings, that is, all projects were considered equally fundable but the primary criterion was the comparative degree of distress. The other projects were held over to later rounds; some of these may have been later withdrawn because of timing problems.

INCREASED DEMAND FOR FUNDING

In 1980, there was an unexpected surge of large city applications; HUD anticipated that the higher demand for program funds might be permanent and examined ways to deal with it. The major approach considered at that time was to reduce the number of applications by tightening eligibility requirements; in particular, by raising the number of eligibility points from three to four. However, it was determined that, even though a four-point requirement would substantially reduce the number of eligible cities, it would have little effect on the volume of applications—the less distressed cities that would be affected by the change were not the cities applying for the program. In any case, the "problem" disappeared on its own.

This same "problem" has recently arisen, but now for both large and small cities. The demand for funds from fundable projects has exceeded the available supply and, once again, the situation is anticipated to continue. The increase in large city fundable projects may be the result of improvements in the economy, some cities' increased capacity to develop "fundables," or to other factors. The large increase in small city projects may likely be the result of HUD's response to what was considered, until recently, *the* small city problem—inability to use the 25 percent set-aside.

Since the beginning of the program, small city awards fell short of the funds appropriated. By the end of FY 1982, HUD had awarded only 20 percent of the small cities UDAG allocation rather than the minimum 25 percent. Unused funds carry over to the next fiscal year together with any de-obligated funds from project cancellations and terminations. Attempts in the past by HUD to reduce the set-aside below 25 percent were met with strong congressional opposition. Most recently, a HUD legislative proposal to shift 117 central cities with less than 50,000 from the large city to the small city pool was rejected. By the end of FY 1982, the small city surplus reached $126 million (compared to the $110 million annual level of funding provided by the set-aside). In response to the surplus, the House Appropriations Committee cut $100 million from the FY 1983 budget, thereby reducing new small city funding to $10 million. However, the Conference Committee restored the $100 million because of the expressed concern that a $340 million level for the program would become permanent. The next attack on the surplus came from OBM that, later in 1982, directed that the FY 1984 budget show a carryover of $244 million in FY 1982 funds to FY 1984, of which $126 million was associated with the small cities program. However, in a rider to the Jobs Bill, the House cancelled the deferral, restoring FY 1984 funds to the $440 million level.

Congress attributed the small city "problem" to insufficient technical assistance and outreach and to UDAG staff cuts affecting application processing. The GAO also examined the low participation rates of small cities and blamed them on lack of knowledge of the program, inadequate capacity to prepare an application, and difficulty of obtaining private sector involvement. These findings are not surprising since 79 percent of the eligible small cities have populations below 2,500 (and 40 percent have populations below 500). HUD's only position at the time was that low participation and the surplus stemmed from the state of the economy.

However, in response to the budget attacks, HUD undertook efforts to increase participation through outreach activities to small cities and private sector participants, increased training and designation of field economic development staff to provide technical assistance, and services of technical assistance contractors. In FY 1983, there was a substantial increase in small city applications, whether due to improvements in the economy or HUD efforts, such that almost $100 million more than the annual appropriation was awarded. The flow of applications did not abate and, in addition, a backlog of held-over fundable applications developed.

In contrast to the thinking in 1980, which considered addressing excess demand by cutting back eligibility, HUD is responding to the issue in 1984 by developing a methodology for ranking fundable projects that will provide a nondiscretionary way to choose among them. The approach, for the first time, explicitly addresses all of the selection factors found in the legislation and program regulations—the latter some seventeen factors addressing both direct program concerns such as UDAG dollars per job and other public policy concerns such as energy conservation.

Under the new system, projects that are technically fundable are ranked on a composite score built around three main categories described in the statute—the impaction measure described above, another distress measure based on unemployment, job growth/lag, and per capita income, and a measure of project impact. The impaction measure is weighted 40 percent because the statute defines it as the primary criterion. The other two categories of measures are each weighted 30 percent. Among the 30 percent for project impact, the leverage ratio and grant dollars per job are counted most heavily, followed by some fifteen other factors that share the remainder. These other factors include total jobs, percentage of minority jobs, local tax collection potential, construction job creation, minority business participation, and so forth.

One issue still to be resolved concerns the small city distress measure. Small city data are not available for some distress factors used for large cities. In addition, HUD wishes to first study the applicability and meaning of extending assumptions regarding indicators of large city needs to small cities.[6] Currently, the small city scoring system for distress measures has been adapted as closely as possible to the large city approach, while HUD reviews this issue further.

Perhaps the most striking features of the new HUD approach are its quite literal application of detailed regulatory criteria and the absence of room for secretarial discretion. The system connects all project selection factors to point scores and ranks projects on their total. Secretarial discretion largely, then, applies only to how a down the list funding will occur, that is, how much of the annual appropriation to award each quarter. Because the statute places predominant emphasis on city distress measures rather than project impact, secretarial flexibility in choosing among otherwise fundable projects is, however, largely constrained in any event.

TARGETING TO PROJECT TYPES

REASONABLE BALANCE REQUIREMENT

The UDAG program administratively classifies projects into three types: industrial, commercial, and neighborhood. This classification scheme was developed to conform to the 1977 Act's requirement that there be a reasonable balance among projects that are designated to restore seriously deteriorated neighborhoods, reclaim for industrial purposes underutilized real property, and renew commercial employment centers. The reasonable balance requirement was considered at the time as a means of encouraging neighborhood projects and neighborhood group involvement and of overcoming a competitive disadvantage such projects might have in attracting private investment for small commercial and housing projects in seriously distressed neighborhoods.

The program office implicitly interpreted the balance requirement only in terms of the number of projects in each category. From the beginning, the program experienced difficulty in attracting enough viable neighborhood applications. Local government preferences were for industrial and larger scale commercial developments. Small cities submitted industrial applications; large cities submitted commercial applications. To increase the neighborhood numbers, contacts were made with neighborhood groups and technical assistance was provided. Projects were classified as neighborhood if they were predominantly housing (wherever located), commercial outside the central business district, sponsored by a neighorhood-based organization (whatever the activity or location), or if they somehow benefited residents of a specific "neighborhood." HUD publicized that neighborhood applications would only compete with each other and not against commercial or industrial applications. Cities, aware of the funding advantage of the neighborhood label, tailored projects that could be justified being called "neighborhood." By the end of FY

1980, the program had awarded enough neighborhood projects to attain its definition of balance. Then, in 1981, the Omnibus Reconciliation Act removed the balance requirement.

EFFECT OF THE 1981 AND 1983 ACTS

By the end of FY 1981, neighborhood projects accounted for slightly over one-third of all projects. In FY 1982, such projects received about one-fourth of the awards; in FY 1983, less than one-fifth. Efforts were shifted to generate more small city applications of all types. HUD not only stopped encouraging neighborhood projects, but also took steps to discourage applications for the major component of the neighborhood classification—housing. The 1981 Act stressed economic development and job creation, and housing projects did not create jobs. However, housing applications continued to be submitted and the projects continued receiving awards in FY 1982 and FY 1983, actually increasing the number of units receiving assistance over prior years.

In 1983, HUD drafted regulations that would prohibit projects that were purely housing, allowing housing units only in mixed-development projects, and only if closely related to job creation. But there was strong congressional opposition to the draft regulations. The House version of the HUD Authorization Bill not only restored the balance requirement but stated it in terms of "amounts available for grants," that is, a dollar balance. The House version also prohibited HUD from denying assistance "on the basis that such assistance is to be used solely for the provision of housing." Part of the opposition was related to cut-backs in other HUD housing programs, particularly those involving production. HUD withdrew the draft regulations. The 1983 Act that passed only included a provision that HUD "may not discriminate among programs on the basis of the particular type of activity involved, whether such activity is primarily a neighborhood, industrial, or commercial activity."

DISTRIBUTION BY PROJECT TYPE AND CITY SIZE

Following the elimination of the reasonable balance requirement in the 1981 Act, proportions of commercial projects in large cities and industrial projects in small cities increased substantially. Perhaps some of these projects had neighborhood features and would have been so classified in previous years. In the four funding rounds of FY 1983, commercial projects accounted for almost two-thirds of large city UDAG funds, and, in small cities, industrial projects received over half of the funds.

TABLE 6.1   Percentage Distribution of UDAG Awards: Small City, Large City, and Total by Project Type, FY 1978-FY 1983

	Small City		Large City		Total	
Project Type	Number	Dollars	Number	Dollars	Number	Dollars
Industrial	44	53	24	17	33	25
Commercial	31	30	44	59	38	52
Neighborhood	25	17	32	24	29	23

The percentage distribution of cumulative FY 1978 to FY 1983 projects and dollars, by city and project type, is provided in Table 6.1.

Overall, commercial projects are the dominant type with 38 percent of the projects and 52 percent of the dollars. The dominance of industrial projects in small cities and commercial projects in large cities is also apparent in the table.

HOUSING PROJECTS

In spite of the drop in projects under the neighborhood classification and HUD's attempt to limit housing UDAGs, the number of units receiving awards in FY 1982 and FY 1983 continued to increase over the level in FY 1981. There were many of these projects in the pipeline, and HUD could not legally stop housing applications under the existing statute and regulations anyway. Neighborhood projects became almost entirely housing related by FY 1983. In FY 1981, 65 percent of these projects involved some housing; by FY 1983, 98 percent of them involved housing.

Since the beginning of the program, awards have gone to projects with more than 86,000 planned housing units. Such projects have been very popular with some cities—38 of Baltimore's 54 UDAG awards have included a housing component. While most housing units are found in neighborhood projects, commercial projects accounted for about 12 percent of the units, and even some industrial projects included housing. A little more than half are rehabilitated units; the remainder are newly constructed. However, there has been a major shift in the split between rehabilitation and new construction in the past fiscal year. Until FY 1983, rehabilitation exceeded new construction except in UDAG's first year. In FY 1981 and FY 1982, around two-thirds of the units involved rehabilitation. In FY 1983, 73 percent of the more than 16,000 units involved new construction, making the UDAG program HUD's second largest production program, with only the Section 202 Elderly Housing program exceeding it.

Overall, 39 percent of the new and rehabilitated units have been planned for low- and moderate-income families. This proportion has declined steadily since the first year of the program, from a high of 65 percent down to only 21 percent in FY 1983.

PROJECT FUNDING SUCCESS RATES AND TERMINATIONS

As indicated previously, the success rate (percentage funded) for all applications has been 50 percent, with large city projects being funded at a rate of 56 percent and small cities, 44 percent. Some variation is found among projects and their locations in large or small cities. Overall, commercial projects had the lowest success rate, 45 percent, compared to 53 percent for both neighborhood and industrial projects. However, this lower rate was associated entirely with small city commercial projects that were funded at a rate of 39 percent. Within large cities, neighborhood applications had the lead (63 percent success), followed closely by the industrial category (59 percent); commercial projects still managed a 50 percent success rate. Within small cities, industrial projects were funded at a 50 percent rate, neighborhood projects, 43 percent, followed by applications with the lowest rate—commercial.

Little variation among project types occurs in the rate of termination. Projects may be terminated by "cancellation" (before grant agreement execution), "mutual convenience" (continuation is agreed to be infeasible), or "for cause" (default). Overall, 11 percent of the projects receiving awards through the end of FY 1983 were terminated. Only 10 percent of the terminations were "for cause." There was very slight variation in the termination rate between large and small cities (11 percent and 10 percent, respectively) or between industrial, commercial, and neighborhood projects (12 percent, 11 percent, and 9 percent, respectively).

ASSESSING PROGRAM BENEFITS

An assessment of the benefits from the UDAG program requires several stages—a determination of just what it is to which impacts should be attributed (e.g., was the grant really necessary in the first place to induce the results), measurement of what specific project benefits were and whether they met project expectations, and identification of broader benefits (such as general city revitalization, spin-off development in depressed areas of cities, or overall enhanced capacity of distressed cities to respond to their own needs).[7]

LEVERAGING PRIVATE INVESTMENT

The key concept in understanding program benefits is leveraging, in which the Action Grant is intended to stimulate private investment in a distressed community, with the added provision that the invest- ment would not occur in that location otherwise. It is not enough that the grant stimulate *greater* private investment in a distressed area than would have occurred otherwise, though this too could generate net increases in job and other benefits for the area (McDonald, 1983). The investment decision must have been such that development would not have occurred there at all without the UDAG.

Leveraging may be considered a proxy for the total public benefits flowing from the program, if assumptions regarding the essential role of the grant are correct. The ratio of the private investment to the grant then can be thought of as an indicator of how much private activity (resulting in turn in public benefits such as jobs, taxes, and improved climate for further development) is actually being produced by each UDAG dollar.

The leverage ratio in UDAG has two meanings—one for indi- vidual projects and one for programwide. The programwide ratio is based on aggregate private investment and total grant funds for the period being considered. The individual ratio, of course, is based on single project data.

On the whole, the programwide leverage ratio is higher than a typical project ratio. The overall program ratio is 5.8 to 1; that is, for each dollar of UDAG, 5.8 dollars of private funds are being invested in distressed cities. The median project ratio is 3.9. One difference between the programwide average and the individual project median occurs statistically. No project, by program rule, is allowed to drop below a ratio of 2.5 to 1, but some have ratios that are high— approaching 20 to 1. The few projects with high private investment relative to the UDAG grant pull the programwide averages up.

Another difference between the ratios occurs because of the way private investment is calculated. For purposes of the individual project leverage ratio, private investment includes all funds commit- ted to the project from private sources *plus* the present value of lease payments, the present value of the UDAG loan if it is repaid at a fixed rate and term, and the present value of loans made to the project from public sources, such as the Small Business Administration or state governments. Industrial Revenue Bonds (IRBs) are also included in both ratios at face value.

The UDAG loan, lease payments, and other public agency loans are not included in the total private investment actually reported for projects when summing for programwide calculation. The effect of having them in the project leverage ratio is to add an incentive to use the UDAG funds as a loan rather than a grant since they improve the leverage ratio for the project and thus may somewhat improve an application's competitive position in tight funding rounds. In cases of some highly marginal projects, it may make the difference for a project in meeting the minimum ratio of 2.5 to 1.

Leverage ratios differ appreciably among types of projects. The program average for industrial projects tends to be higher (about 7 to 1), and neighborhood projects lower (about 4.5 to 1). Commercial projects hold close to the programwide average—about 5.7 to 1. Small city project ratios tend to be greater than large city—5.9 compared to 5.6. This occurs partly because small city projects are more often industrial.

Over the history of the program, there is some evidence of leverage ratio "deflation," most evident for FY 1983, with ratios dropping from over 6 to 1 in early program years to 5 to 1. This is probably attributable to several factors, including tighter economic conditions in distressed cities that make the development "gap" greater, and perhaps, in contrast to the early years of the program, reduced availability of vacant urban renewal land and readily developable sites.

In its most commonly used forms described above, the programwide leverage ratio ignores some of the financing commonly involved in a project. Projects frequently include other federal, state, and local funds as well as the UDAG and the subsidized and unsubsidized private funds. These various components can be combined in ratios in a number of ways, and doing so can change the picture of a project. One alternative is the ratio of private funds to all public investment, not just to the UDAG. This approach assumes that all of the public funds are necessary to leverage the private investment and perhaps gives more of a picture of what the total subsidy actually is in a project. By this ratio of private investment to total public funds, the programwide average for UDAG is 3.7 to 1 (4.8 for industrial projects, 3.5 for commercial, and 3.0 for neighborhood).

Another possible ratio is that of private plus state and local funds to the UDAG dollars. This approach assumes that the UDAG funds were necessary to stimulate the nonfederal expenditure, and gives the largest value of the alternatives. By this measure, the program has leveraged 6.1 nonfederal dollars per dollar of UDAG. A refinement is to add all other federal expenditures in the denominator, resulting in a 5.3 to 1 programwide ratio.

FEASIBILITY AND SUBSTITUTION

A sidelight to assessing program benefits is that a project must be feasible before it is realistic to expect that any of the public benefits will flow from the proposed investment. While this sounds like a truism, it poses one of the most challenging tests to the program staff underwriting the deals and a tricky situation to the developer. As Munkacy and Rappaport put it:

> The "Catch 22" of UDAG is that, for developers to prove a UDAG is needed, they must first prove that it is not needed. That is, the development strategy, concept, and program must be rooted in strong market support and must capture market potentials [Munkacy and Rappaport, 1983].

The UDAG program must walk a narrow line between the risks of financing economic activity that is unlikely to succeed, and subsidizing investment that would have occurred anyway. The first problem has its own built-in indicator, at least for hindsight—whether the project will sooner or later fail. The second is far more difficult to diagnose. Even if it can be established that the amount of funds represented by the Action Grant is needed for the development to succeed, the UDAG program still may not have been needed to supply those funds. This is the so-called substitution issue. Substitution occurs if the UDAG is unnecessary and other federal, state and local, or private funds would have been used to complete the deal in the absence of UDAG. In the UDAG program, substitution is illegal. In late 1979, the program authorizing legislation was amended by the so-called Wydler amendment to require that UDAG funds not substitute for or replace other nonfederal funds.

Both reviewing projects for evidence of substitution before funding approval and evaluating the program for evidence of it after the fact are extremely difficult tasks. While much talked about, the evaluations have rarely been actually undertaken. Several approaches have been proposed. One has been to analyze net national investment to assess whether the UDAG program has contributed to an increase in overall economic activity nationwide or has just substituted for private investment that would otherwise have taken place somewhere. While never actually carried out (and probably impossible to do at the national economic level), this thinking appears to have guided some proposals to abolish the program on the presumption that the analysis would show pure substitution. This conception of substitution, however, misunderstands the UDAG concept, which is to influence *where* investment takes place, not *whether* it takes place.

A more appropriate approach to the substitution question might be to use the same methodology to analyze net community investment in distressed cities, either on a time series basis in participating UDAG cities or on a cross-section basis comparing participants and nonparticipant cities, or both. To our knowledge, this has not systematically been attempted, and it presents a constructive research possibility.

The one approach that has been attempted is a case-by-case review of actual Action Grant projects by real estate finance and development experts. This was done in the evaluation of the UDAG program conducted by the Office of Policy Development and Research in HUD in January 1982 (HUD, 1982). Eighty randomly selected projects were reviewed by a panel of experts to assess whether the project would have had the same scope, operated in the same city, and occurred at the same time without the UDAG. The panel used information from project files and from interviews on-site to reach their conclusions. Full substitutions was defined as existing if all three conditions occurred; partial substitution if the scope would have been different.

The study found evidence of full substitution in 8 percent of the projects—in the opinion of the reviewers, these projects would have clearly gone ahead at the same time, place, and scope. Of the projects, 13 percent were found to have partial substitution and another 15 percent were inconclusive. No evidence of substitution was found in 64 percent of the projects. However, it should be noted that some of the projects reviewed in the HUD study were approved before the Wydler amendment.

On the basis of these findings, the Action Grant program reviewed its underwriting and documentation practices—and revised its application procedures—to strengthen feasibility reviews and tighten declarations of need for the Action Grant. While there has not been a subsequent evaluation of the substitution question, it is likely that its frequency as defined by the study has declined since the HUD evaluation.

PROJECT IMPACTS

In the most immediate sense, project impacts are measured in terms of jobs and local fiscal revenues. These are the two categories of impact (along with leveraging) cited in the statute as criteria for grant selection (housing is not mentioned at all), and are the two that have been most frequently emphasized by the program.

Job creation. Expected job impacts are categorized in a number of ways, such as total new permanent jobs per project, construction job creation, job retention, and UDAG dollars per new permanent job. The latter figure is probably most often relied upon in the program as an indicator of the job impact of a project. Jobs for low- and moderate-income persons are also considered important. These jobs are defined by the program in terms of accessibility to persons from families with low- and moderate-incomes. In practice, these jobs are probably those that pay low- and moderate-income wages. Jobs to be filled by minorities are also identified in projects, as well as those to be filled by CETA eligible people, as required by statute. The CETA provision has been maintained in spite of the demise of the CETA program, and CETA eligible workers are defined according to the regulations of the former program.

Total expected new permanent job creation estimated by the UDAG program through FY 1983 is about 411,000 jobs. A little over one-half (55 percent) are jobs for low- and moderate-income people. Most jobs are created in commercial projects (56 percent), and individual commercial projects tend to average more jobs per project than other types (304). Industrial activities average 204 per project. Most planned UDAG jobs, of course, are in large cities. Expected employment creation in large cities is 294,000 jobs (just under 1 percent of the total 1980 labor force in the cities receiving awards).

UDAG projects also include job retention in some circumstances, though reported numbers of such jobs have dropped dramatically over the program years—from 35,000 in FY 1979 to under 7,000 in FY 1983. This drop is mostly a result of a greater rigor in counting retained jobs actually attributable to the program.

Estimates of construction jobs have also been refined and are now calculated on an annual equivalent basis by computing the labor cost of construction, drawn from the percentage of construction cost in the project attributable to labor (using contractors estimates, national industry standards, or other techniques), and dividing this by the average annual construction salary and fringe benefits. Generally, construction job estimates for projects approved before this estimation technique was adopted have been corrected or deflated comparably, and, overall, the program estimates that more than 311,000 annual equivalent construction jobs are being created by the approved projects.

One important question, of course, is just how many of the expected job goals are actually achieved. One way is which reality can fall short of expectations is if approved projects are terminated. As

indicated earlier, 11 percent of the approved projects have been terminated. The planned job figures reported above are net of these terminations. Terminations annually affect about 7,000 to 10,000 planned
jobs, though funds from terminated projects are available for future
approvals.

Job estimates also can be wrong because original projections were
inaccurate. The one systematic effort to assess whether projects are
actually achieving original hiring goals is the HUD evaluation. This
study concluded that the projects examined could be expected to
achieve 77 percent of the jobs estimated at the time of application.
The shortfall resulted from errors in estimation in some projects and,
in others, from adverse economic conditions. Since the HUD report
was released in January 1982, the UDAG program has substantially
revised the way in which job estimates are made—converting all
calculations to full-time equivalents, broadening the supporting
documentation required, and in some cases simply proportionately
reducing all estimates for certain types of projects that traditionally
involve large amounts of job transfer (such as office work). As a
result, recent new permanent expected job estimates may be closer to
levels that will actually be achieved than was the case at the time of
the HUD study.

UDAG job estimates through the end of FY 1983 indicate that the
average UDAG dollars per job is $7,432. This has increased annually
(because of inflation and because job estimation techniques have been
tightened), starting at $5,700 in FY 1978 and now amounting to
$9,500. Commercial jobs are slightly more expensive than industrial,
averaging $6,896 for commercial and $5,821 for industrial. Neighborhood projects are much higher than either of these others—$14,412—
and pull the total program average up.

Because of the many complexities in measuring expected job
creation, it is difficult to compare job creation costs across federal
programs. Disparities exist in what jobs are counted, over what
period the jobs are considered, and whether multiplier effects are
included or not, as well as in how the corresponding program expenditures are calculated—especially whether substitution is considered.
One comparison has been made between UDAG and the EDA
Business Development program based on roughly comparable program characteristics, which found the UDAG program to compare
favorably at about 88 percent of the cost per job in the EDA program
(HUD, 1982). Another available comparison is between the UDAG
and CDBG, based on estimates reported in the 1983 Annual Report
on Community Planning and Development programs. While not

strictly the same in purpose or measurement, this comparison is striking: UDAG is about one-third of the estimated $30,000 per job for CDBG.

TAX REVENUES

Many observers would conclude that projecting tax revenues from complex urban development deals defies human capacity. The HUD study found that tax revenues actually generated by UDAG projects fall short of projections made at project initiation by as much as 50 percent. This is undoubtedly in part a result of speculative and generous estimates at project outset (unlike job goals, tax revenue increases are not usually made a part of the grant agreement between HUD and the city and thus are not as rigorously reviewed), as well as other features of the tax estimation process, such as including taxes not paid directly to the city and averaging effects of the abatements over twenty years. In any case, again in response to the HUD findings, the UDAG program has greatly tightened up its process for estimating expected tax revenues. The application now includes three pages of instructions and three pages of forms for this estimate and requires a written certification of the estimate from the local chief tax assessor or fiscal officer.

Unadjusted for any overestimation, total expected annual tax revenue increases for all projects approved under the program is $469 million. Of this expected increase, 64 percent comes from local property tax increases, 32 percent from other tax revenues, and 4 percent from payments made in lieu of taxes (PILOTs).

RECAPTURE

An approach in the UDAG program that has potential for altering direct city benefits is the use of the UDAG as a loan to developers, to be paid back to the city for future development projects. In FY 1978 and FY 1979, only one quarter of the projects involved some kind of payback to the community; since that time, over three-quarters of the deals have been structured to provide paybacks.

Clarke and Rich (1982) have described this process as the emergence of a group of "entrepreneurial" cities that are able to "make money" on their grant. They identify several major mechanisms through which it takes place: land disposition by which program income is generated by sale of land to the private sector; lease agreements and municipal enterprises, for example, with local parking facilities; loans in which developers pay back grant funds at a fixed rate and term or with a sliding interest scale; and participation in equity or net cash flow.

However, the open question is how much money cities actually make. Many loans and other payback schemes are soft—second or third mortgages or shares of "excess" profits after payment of syndication proceeds, and so on. The "entrepreneurial" approach appears to provide many positive incentives to all the participants in the deal, but it will be many years before it is clear whether it makes money for cities. So far, reported paybacks in the program are a cumulative total of $44 million.

URBAN REVITALIZATION

But net project jobs, taxes, and paybacks from each project are probably not enough alone to assess the impact of the UDAG. The concept (and the statute) intend not just for UDAG dollars to leverage private dollars in a project, but for UDAG projects to leverage or stimulate further economic revitalization or economic recovery of distressed areas. There is no question that UDAG funding has made the difference for many urban development initiatives in distressed areas, and support for it by mayors and development directors is strong. The real test now, and perhaps the key emerging evaluation question for the program, is whether successful individual projects in distressed cities can in turn stimulate or help sustain spin-off investment and local capacity for self-development necessary to continue revitalization.

APPENDIX: Urban Development Action Grant Program Information, FY 1978-FY 1983

	Small	($Millions)	Large	($Millions)	Total	($Millions)
Number of projects	871		1,101		1,972	
Neighborhood	218	124	351	571	569	695
Industrial	385	382	262	389	647	771
Commercial	268	215	488	1,372	756	1,588
Number of cities	635		294		929	
Leverage ratio	5.9		5.6		5.7	
Action grants		721		2,332		3,054
Private committed		4,276		13,228		17,504
Other federal		102		348		450
State/local		234		960		1,195
Total investment		5,334		16,869		22,203
Tax rev. increase		84		370		454
Tax rev. per AG $	.09		.10		.10	
AG $ per job	6,173		7,933		7,432	
New permanent jobs	116,852		294,006		410,858	
Low-income jobs	69,638		158,550		228,188	
Retained jobs	27,335		89,797		117,132	
All housing	11,865		75,023		86,888	
New housing	8,008		32,676		40,684	
Rehab. housing	3,857		42,347		46,204	
Low-income housing	4,405		29,704		34,109	

SOURCE: HUD (1983).

NOTE: Data are derived from applications announced for preliminary approval, excluding terminated projects, and are the anticipated results when all projects are completed.

NOTES

1. For a detailed, documented review of the development of the program in its first two years, see Greene (1980).

2. For a comparison of UDAG and enterprise zone proposals, see Green and Brintnall (1983). Also see Brintnall (forthcoming).

3. Unless otherwise indicated, data in this chapter are drawn from HUD (1983) or from the Office of Urban Development Action Grants. The appendix provides program summary data for fiscal years 1978 through 1983.

4. For a discussion of why these measures were selected, and why others were rejected, see GAO (1980). For an analysis of the pros and cons of each measure, see HUD (1979).

5. A full list and description of the selection factors is in the *Federal Register* (Vol. 47, No. 36, Feb. 23, 1982).

6. The variable of greatest concern is perhaps the measure of percentage of housing built before 1940. This measure was devised as an indicator of infrastructure need in large cities—and is firmly embedded in many large city need measures, including the CDBG entitlement city formula. Its basis as a small city need measure, however, has not been established, although it has readily been applied to small cities in the past.

7. For a discussion of the conceptual issues and measurement problems in a comprehensive framework for evaluating federal economic development investments, see Redburn et al. (1984).

REFERENCES

BRINTNALL, M. (forthcoming) "Beyond white collar crime: public policy and the boundaries of economic exchange." Law and Policy Quarterly.

CLARKE, S. E. and M. J. RICH (1982) "Partnerships for economic development: the UDAG experience." Community Action 1, 4: 51-56.

GAO [U.S General Accounting Office] (1980) Criteria for Participation in the Urban Development Action Grant Program Should Be Refined. Washington, DC: Government Printing Office.

GREEN, R. E. and M. BRINTNALL (1983) "State administered enterprise zones: dimensions and experience." Presented at the annual meeting of the American Society for Public Administration, April.

GREENE, A. (1980) "Urban development action grants: a housing-linked strategy for economic revitalization of depressed urban areas." Wayne Law Review 26, 5: 1469-1504.

HUD [U.S. Department of Housing and Urban Development] (1979) Pockets of Poverty: An Examination of Needs and Options. Washington, DC: Author.

———(1982) An Impact Evaluation of the Urban Development Action Grant Program. Washington, DC: Office of Policy Development of Research.

———(1983) Urban Development Action Grants Data Book, Vol. 30. Washington, DC: HUD, Community Planning and Development, Office of Management and Statistics Division.

McDONALD, J. F. (1983) "An economic analysis of local inducements for business." Journal of Urban Economics 13: 322-336.

MUNKACY, K. and J. RAPPAPORT (1983) "The UDAG Program: alive and well and taking kickers." Urban Land (December): 2-6.
REDBURN, F.S., S. GODWIN, K. PEROFF, and D. SEARS (1984) "Federal economic development investments," in M. Holzer and S. Nagel (eds.) Productivity and Public Policy. Beverly Hills, CA: Sage.

7

Free Zones in the Inner City

STUART M. BUTLER

☐ IN AN ARTICLE in *The Urban Lawyer* early in 1983, David Callies expressed bewilderment at the attention given to a relatively minor piece of urban legislation then making slow progress through the U.S. Congress. The measure in question was designed to create a number of so-called enterprise zones in the heart of America's most distressed cities. Wrote Callies,

> Rarely has the federal government had the dubious benefit of so much advice on a modest piece of proposed legislation that fails to provide a single direct dollar of federal aid. For a bill (or series of bills) that has yet to come up for a vote before either the House or the Senate, the attention lavished upon this legislation is a trifle overwhelming [Callies and Tomashiro, 1983: 232].

The enterprise zone concept[1] has been variously described as creating "mini Hong Kongs" in the nation's most blighted inner cities, "greenlining" these areas, or setting up capitalist enclaves in depressed and stagnant urban communities. The concept rests on the notion that within even blighted neighborhoods there usually exists considerable potential for economic development, but that it is rendered dormant by a tax and regulatory system that suffocates initiative and self-improvement. Moreover, these resources are often ignored even by residents, because government programs tend to require people and places to demonstrate weaknesses, not strengths, if they are to obtain assistance from government.

An enterprise zone would involve reducing tax and regulation in designated inner-city neighborhoods with a view to stimulating risk taking and adaption. The idea would be to generate new ventures and social organizations to rebuild the local economy from the bottom up, instead of the more common practice of outsiders imposing a plan on the neighborhood. History shows that the inherent strength of

American communities and institutions can be attributed in large part to the freedom Americans had to experiment without the deadening hand of central control. Frontier communities were left alone to devise ways of dealing with unusual crises and opportunities without having to conform to rules that bore no relation to local conditions. The enterprise zone concept sees inner cities, in effect, as "urban frontiers," constrained by regulations and a tax code that is irrelevant to the problems and opportunities that exist. By reducing these suffocating barriers, the aim would be to stimulate frontier creativity in urban neighborhoods.

The zone idea runs counter to the theory underpinning the development strategy of most major American cities. Rather than engaging in the zero-sum game of "smokestack chasing" in an attempt to attract or retain large companies, the enterprise zone approach draws on recent evidence on job generation to make the argument that cities should focus instead on stimulating small, new, indigenous enterprises. This would be accomplished by "supply-side" tax and deregulatory policies aimed at establishing a "blind" environment attractive to small entrepreneurs. It is the central thesis of enterprise zone proponents that people in depressed communities are far better able to tackle their economic and social problems, if given the opportunity to do so, than any number of well-financed outsiders.

The enterprise zone proposal, then, has led to an intense debate because it provides a focus for rethinking many key aspects of urban policy. Whether or not the measure eventually passes Congress in a recognizable form—and that will depend more on politics than on the merits of the concept—the concept has raised fundamental questions about the process of economic development in cities.

The term "enterprise zone" was first coined in 1978 by Sir Geoffrey Howe, who in 1979 became Chancellor of the Exchequer in Mrs. Thatcher's Conservative government. A version of the idea was enacted by the British Parliament in 1980. Prior to passage of this legislation, however, the rudiments of the concept were "exported" to the United States, where they gave rise to the discussion and legislative activity still in progress. Tracing the development of the British model and its American offspring provides us with a useful framework in which to analyze the enterprise zone concept and the many issues and questions surrounding the use of low-tax economic zones to combat urban blight.

ENTERPRISE ZONES IN BRITAIN

While the term originated with a conservative politician, the foundations of the concept were provided by Peter Hall, a highly respected socialist academic at Britain's Reading University. Surprisingly for a former chairman of the Fabian Society,[2] Hall (1977) maintained that Britain's seeming inability to emerge from the structural decline of its traditional industrial economy is due in large measure to an overdose of bureaucracy, which has stifled creativity and economic adjustment by discouraging entrepreneurship. The problem is particularly acute in central metropolitan areas, Hall argued. In these communities, residents have been left stranded by migrating industries and by the rapid evolution of technology, rendering their skills obsolete.

Yet, technological change is nothing new, Hall noted. Cities were once the crucible for change because they were havens for adaptable small entrepreneurs; the inner cities of Britain were at one time said to be the workshops of the world. But now, that reputation has passed to the many free trade zones around the world, in which tax and regulatory barriers have been removed in a successful attempt to generate new enterprises.

Hall's conclusion was that Britain should recognize the success of the free trade zone model and utilize it. Rather than vainly trying to predict the unpredictable future shape of the urban economy, some of the most blighted districts of Britain's cities should instead be turned into what Hall termed "freeports." The freeports would be

essentially an essay in non-plan. Small, selected areas of inner cities would be simply thrown open to all kinds of initiative, with minimal control. In other words, we would aim to recreate the Hong Kong of the 1950's and 1960s inside inner Liverpool or inner Glasgow [Hall, 1977].

Hall saw his freeports as providing seedbeds for the rapid cultivation of firms able to adapt to new technologies and to the existing labor pool, which would then give birth to a new generation of industries appropriate to the urban environment. The process of change would be unplanned and unbureaucratic—it would be the trial-and-error process of the marketplace.

Hall's freeports were to be similar to free zones such as Hong Kong. They would be outside British foreign exchange and customs controls, as an inducement to foreign capital and international trade, and personal and corporate tax rates would be reduced to a minimal level. On the other hand, government services would also be reduced to the bare essentials. Residents would have to forgo government in most of its forms. And for that reason, he suggested that the freeports should be located in extremely derelict unpopulated areas—so that people would have to make a conscious decision to move into them.

While the Hall proposal generated considerable debate, it was not taken up by any major politician. Even Conservative Party politicians found the thought of zones incorporating a minimalist state much too radical to contemplate. Hall's freeport did, however, provide the impetus for the proposal that finally became Britain's enterprise zone program. Yet, the Conservative version turned out to be a very diluted form of Hall's "Hong Kong on the Thames."

As it pondered the enterprise zone idea, Thatcher's government noted that it is a peculiarity of British cities that great expanses of valuable commercial land lie idle because its government owners have not permitted the parcels to be developed. In most cases, these sites are in public ownership because they belong to government corporations. The government believed that the key purpose of the legislation should be to bring about the development of such large expanses of derelict or vacant urban land, owned by government in order to stimulate urban job creation.

The final British legislation, passed in November 1980, reflected this view. Cities were invited to submit applications for areas of up to one square mile to be designated as enterprise zones by the national government. Eleven zone "slots" were on offer and there was strong competition between cities for the designations. Only broad guidelines were given to indicate what would constitute a successful application, but there had to be agreement between the city and the national government regarding the boundary of the proposed zone, and it was made clear that priority would be given to sites that would be developed rapidly.

Under the law, the designation lasts for an agreed period, normally ten years, during which the following tax and regulatory changes apply:

(a) *Development Land Tax.* Development Land Tax is a capital gains tax applied to that element of a land or property sale deemed to arise from its potential for future development. The tax is said to inhibit sales of vacant land. Within the enterprise zone, the tax does not apply.

(b) *Property tax.* Property taxes are not levied on any business-related buildings in the enterprise zone, but this relief is not extended to residential property. Under the zone law, all tax revenue lost by a local government—on existing or new structures—is reimbursed to that government from the national treasury.

(c) *Capital allowances.* Generous depreciation schedules—or "capital allowances" as they are called—have operated throughout the United Kingdom for many years. Fifty percent of capital expenditures on the construction or improvement of industrial buildings, for example, is deductible against taxable income in the first year. Capital allowances and capital grants are also available from the national government for projects in certain regions of the country, as an intergral part of a national regional policy.

Within enterprise zones, the full cost of all business buildings and all equipment may be depreciated within one year. While this is a significant improvement in the tax treatment of industrial buildings, it is an even more generous allowance for commercial buildings, since the normal 50 percent allowance is limited to industrial facilities.

(d) *Zoning.* As part of their appreciation for an enterprise zone, the local government had to agree to streamline building codes and zoning restrictions within the area. In practice, cities were required to adopt performance zoning in the enterprise zones. Any structure is given automatic clearance, provided it meets basic health and safety standards.

It can be seen from the nature of the tax changes in the British enterprise zones that the primary aim of the program is the industrial and commercial development of derelict sites. Virtually all the tax relief is related to *physical* redevelopment for business purposes. A company has to undertake construction or improvement before the most significant incentives are triggered—the only exception being property tax relief, which applies to existing structures. There is no relief of any kind for the hiring of labor, nor is there any general reduction of corporate or individual income taxes within the zones—a far cry from conditions within Hong Kong.

One element of Hall's freeport proposal that did find its way into the enterprise zone program was the notion of picking vacant areas, where widespread business activity could occur without disrupting an existing community. The British government never gave any serious consideration to siting an enterprise zone in a populated neighborhood. Virtually all the sites chosen have been unpopulated, derelict areas: either expanses of wasteland, or places dominated by vacant, obsolete factory or dockland buildings.

CRITICISMS OF THE BRITISH ZONES

The general political reaction was favorable. Even those who were skeptical of the whole idea saw little harm in trying something new in places that were the epitome of failure. And many urban officials, with little interest in the philosophy behind the approach, saw the tax incentives in the enterprise zone as a potential shot in the arm to their efforts to attract job-creating companies.

After an initial flush of enthusiasm, however, the business community—especially small business organizations—began to express deep reservations about the program (which in its early days had been touted as the salvation of small urban enterprise). And before the legislation had even passed, *The Economist* (1980) warned that there was a distinct danger that no new jobs or businesses would be created at all—already successful companies would simply be lured into the zones by the prospect of lower taxes.

Many downtown small businessmen claimed that the zones would attract large retailers who would then compete for the same market. The new entrants could be large chain operations, critics maintained, because the zones offered many acres of vacant land and the generous tax incentives could be used by branches of major profitable corporations. Downtown merchants complained that this combination of factors would give these large outlets significant—and unfair—cost advantages.

In theory, of course, there was nothing to stop small enterprises themselves from moving into the zones. But small business organizations pointed out that the tax relief available was of only marginal benefit to a small new firm. Young companies generally rent rather than buy buildings, and so the generous depreciation available on business buildings offered no direct benefit. They also tend to be labor intensive rather than capital intensive—particularly in the rapidly growing service sector. So the extra tax relief available for equipment used in the zones was less than exciting to them. In any case, small business advocates argued, income taxation is no great obstacle for most small new firms. In their early years, companies are generally struggling just to make a profit on which they can pay tax. Business income tax deductions, in these cases, are of little value. The real need is for capital, not tax shelters.

Government officials countered the concerns of small business by suggesting that even if firms rented, the tax incentives available to developers would be passed through to the struggling entrepreneur in the form of lower rents, and the favorable tax treatment of zone machinery would lead to low cost leases.

REVIEW OF THE ZONES

It is just two years since the first British enterprise zones were created. Given the considerable amount of initial reclamation work that had to be completed in many of the sites, the effective life of the zones has been a little over a year in most cases. Nevertheless, sufficient time has elapsed for some initial conclusions to be drawn.

The British government undoubtedly feels that the enterprise zone experiment has been sufficiently successful to expand the program: In November 1982, 13 new sites were announced. However, the track record so far suggests that several of the early concerns expressed by businesses have proven valid.

(a) Business costs. There is little evidence that the favorable tax treatment of plant and equipment in the zones has been passed through to commercial renters. Far from leading to a reduction in rents, rents in the zones have generally moved higher than the level in adjacent sites. Indeed, landholders near the zones have complained that the effect has been to depress their propertly values, making it more difficult for them to lease out or sell space.

(b) Retailing. Several cities, with the agreement of the Thatcher administration, placed limitations on the proportion of retailing permissable in their enterprise zones, chiefly because nonzone retailers complained of damaging and unfair competition. However, it should be pointed out that the peculiarities of the British tax code make the enterprise zone more attractive to retailers than to any other business group.

(c) Job creation. Jobs have certainly appeared in the enterprise zones. Still in doubt, however, is whether these are really "new" jobs, or merely jobs that either existed or would have existed elsewhere in the country. The government made it clear during the passage of the legislation that it did not expect the zones to have a "vacuum or sucking effect." Yet the evidence to date suggests that most of the zone jobs derive from companies that would have expanded in any case, and merely found the zones attractive places in which to execute that decision.

(d) Small firms. The enterprise zones have not become havens for new entrepreneurs. The cost of doing business for start-up firms has not fallen significantly within the zones, and the tax relief package did little to help small firms attract capital.

While it would be unreasonable to write off the British enterprise zone experiment as a failure until enough time has passed for unambiguous patterns to be seen, it is fair to say that so far the initial promise has not been fulfilled. And the experiment leads us to ask some important questions of the whole strategy of using "economic zones" as instruments to foster economic growth.

Unless one subscribes to Milton Friedman's dictum that any tax reduction is by definition a good thing, no matter who is the beneficiary or what is the pretext, it is hard to justify an economic zone if it merely creates a tax differential leading to the artificial and arbitrary redistribution of economic activity from one place to another, such that resources are used less economically than before.

The British enterprise zones do seem to fall foul of this objection. But it might be argued that free trade zones such as Hong Kong or Taiwan also benefit from artificial differences between their economic environment and those in neighboring countries. But if neighboring countries do not copy Hong Kong and other free zones, that is their problem. Indeed, just like the appearance in the market of an efficient new product, an economic zone may have the effect of pushing nearby governments to adopt policies more conducive to trade and business, generally—to the benefit of everyone in the region. Hong Kong could certainly claim to have had that effect even on its immediate neighbor.

However, the creation by a country of special zones within its frontiers is different from the Hong Kong case. The British enterprise zones involve a government giving a tax advantage to business in a certain place, in the full knowledge that these companies will thereby prosper compared with other firms. Unlike the case of Hong Kong, the British government's role in the enterprise zone is not like one business gaining an advantage because of the ignorance or lack of foresight of his competitors. It is more similar to a cartel granting favorable terms to a few chosen customers.

A domestic zone of this type, however, might be justified if it encourages new economic activity from resources within the zone, without imposing significant costs on the rest of the country. Let us assume, for example, that an enterprise zone was to be designated in a blighted inner-city neighborhood where there existed dormant factors of production. Let us assume that the area had been receiving substantial financial aid (welfare, housing subsidies, etc.) derived from taxation levied on the rest of the country, and also that the principal reason for the plight of the neighborhood was that existing regulation and rates of taxation discouraged utilization of the area's human and other resources. Let us finally assume that the tax and regulatory relief provided within the zone had the effect of stimulating the utilization of the area's dormant resources—without the importation of any resources from the outside area—and that this led both to new tax revenue being generated within the zone and to a reduction in welfare assistance to the zone's residents.

In this (admittedly perfect) case, the effect of the tax and regulatory differential favoring the zone is to generate income for residents out of previously unused resources, and to do so while *reducing* the financial burden on people outside the zone. A zone of this kind is most beneficial to its residents, but it is in enlightened self-interest of other citizens to agree to its creation. The latter group may benefit only marginally from the new activity within the zone, but if they were to insist on the restoration of equal taxation, it would result in a cost to themselves.

While such a perfect case would never be attained, it is quite conceivable that with appropriate relief the cost of an enterprise zone (including capital and other factors of production drawn from other places)[3] would be more than offset by new tax revenues and a reduction in government expenditures within the area. If this is to be achieved, the key requirement is that as much as possible of the activity within the zone must be genuinely *new* in nature—that is, occurring only because of the existence of the zone.

ENTERPRISE ZONES:
THE AMERICAN VARIANT

The enterprise zone concept began to attract interest in the United States in 1979, a few months after Sir Geoffrey Howe had first introduced the term. In Britain, the concept was turned into legislation relatively quickly, with little public debate. In America, almost the opposite has been true. For the last three years, hardly a single conference on urban development has taken place without a lively discussion on enterprise zones. And a considerable body of literature has grown up around the various themes of the idea. Yet, legislative action has been slow. Presidential Candidate Ronald Reagan made a firm commitment to the proposal during the 1980 election, and legislation has been before Congress for four years. But for various political reasons, the zone bill has so far not received the level of White House political backing needed for passage. Meanwhile, almost half the states have passed legislation to create enterprise zones, within which relief is given from state and local taxes and regulations. Several states have already designated their zones.

The debate on enterprise zones in America has been thorough. It remains to be seen as to what degree the final combination of state and federal measures will reflect the main points of agreement in this debate. But it does appear that the American version of enterprise zones will be closer to meeting the test outlined earlier than can be said of the British version. The number of reasons for this cautious optimism are as follows:

SITING AND PURPOSE

Some advocates of enterprise zone legislation in the United States share the British view that the only objective should be the creation of economic activity, particularly jobs, in selected urban locations—without regard to whether this is new or relocated activity. But a broad consensus contends that the enterprise zone must be an instrument to revitalize depressed urban neighborhoods by stimulating indigenous enterprise. This view stems in large part from the feeling that the zones offer a strategy to deal with highly distressed and troubled minority areas.

The federal legislation, and most state laws, reflect this commitment to populated areas. Minimum population figures are stipulated in the federal enterprise zone bill, and in many of the state measures, the tax incentives are contingent upon the employer hiring a certain number of zone residents. In some states, there are also provisions to help the formation of locally owned enterprises in the zones (resident-owned cooperatives are specifically encouraged in certain states).

The objective of these provisions is to restrict tax relief as far as possible to enterprises that utilize existing factors of production in what are now dormant areas. Needless to say, no such attempt will be watertight, but there is a conscious attempt to create genuinely new economic activity within the zones; despite its rhetoric, the British government did not even try to do that.

THE IMPORTANCE OF SMALL BUSINESS

The role of small new business has become a central issue in the American discussion of enterprise zones. Again, this was not the case in Britain. One reason for this difference has been the pioneering work of David Birch at MIT, who demonstrated both that young small firms are the most effective generators of new jobs in the economy and that job losses in declining urban areas are due primarily to a low start-up rate of small new companies, not the contraction or shutdown of existing firms (Birch, 1979; 1980). Another reason for the interest in small firms is that in line with the theme of mobilizing indigenous resources within a depressed community, proponents of the enterprise zone are forced to emphasize businesses that could be started by people of modest income in facilities that are already available. Consequently, the small business in a garage or abandoned basement is clearly more relevant to the discussion than is the Fortune 500 corporation. The emphasis on small firms has been further reinforced by the work of urban writers such as Jane Jacobs, who has noted the key social importance of small enterprises to the economics and social fabric of poor neighborhoods (Jacobs, 1961; 1969).

The impediments to small business creation have thus attracted more interest in zone discussions here than in Britain. This has led to pressure for the inclusion of two key elements in the American zones: local deregulation and tax changes designed to meet small business requirements.

Local regulations such as zoning, building codes, occupation licensing, and other restrictions of business formation have long been the target of small businessmen, who argue that "equal" regulation has unequal effects on smaller firms. Moreover, certain regulations that are said to impart benefits to residents and owners in affluent neighborhoods (such as zoning, to maintain property values), have the opposite effect in low-income neighborhoods (where zoning restrictions, for instance, discourage mixed use and may erode community stability). This has led most state enterprise zone laws and the proposed federal legislation to require to streamlining of local regulation as a means of encouraging small firm formation.

Just as a regulation may have different effects on small firms compared with large firms, equal tax relief can also have different consequences. As small business advocates argued in Britain, generous depreciation allowances, or corporate tax reductions, are of limited use to the typical young small firm struggling to register a taxable profit. Far more important are payroll taxes and taxes that affect the entrepreneur's ability to raise capital (such as capital gains tax and the personal income tax).

In the United States, there has been considerable pressure on both states and the federal government to provide tax relief in enterprise zones that addresses the concerns of small start-up companies. The results have been mixed. Attempts in Congress to reduce social security payroll taxes in the federal legislation were unsuccessful. A provision to exempt enterprise zone investments from capital gains tax, on the other hand, remains in the bill. There is a significant tax credit for expanding the payroll and for hiring disadvantaged workers. But these payroll incentives are only available against business income taxes and so are of generally marginal benefit to young small firms. Moves to include tax incentives for tax payers taking an equity share of a new, small enterprise zone company have so far been resisted by the Congress. Some state measures have included relief designed to spur investment in new firms, but these, like the federal incentives, are weaker than might be expected in view of the widely agreed importance of new enterprises.

RELOCATON

The issue of relocation has loomed much larger in the theory and politics of enterprise zones in America than in Britain. Given the checkered experience of American cities in using tax incentives to lure firms across metropolitan boundaries, politicians are concerned lest an enterprise zone merely causes further relocation and tax loss.

Although the federal tax incentives currently under discussion seem most attractive to profitable firms that branch or relocate into the zone, there are reasons to suppose that relocation will be less of a problem than many zone critics fear. If the zones are sited in blighted neighborhoods with high crime and social problems, it would take a substantial tax incentive to induce branching or a relocation. But several studies have shown that firms (particularly large firms) tend to be far more sensitive to a community's social conditions and labor resources when considering a location decision than they are to tax incentives.[4] The British zones, in contrast, are sited in areas that are not known for social distress or unrest. Moreover, the tax relief available is heavy and immediate.

In addition, the principal federal labor incentives in the United States are skewed toward firms that expand their payroll. This does mean the zones would still be attractive to branches of large companies, but they would be of less interest to a firm contemplating simply a relocation of its operations to a zone. Some states have further discouraged relocation by enacting tight eligibility criteria designed to exclude relocating companies from the tax benefits.

CONCLUSION

The enterprise zone proposal is a fascinating attempt to use the mechanism of free economic zones to bring new economic life to highly blighted urban neighborhoods. Peter Hall's freeport idea sought to do this by creating, in effect, independent city-states in major metropolitan areas. The subsequent British enterprise zone program is different but has provided two chastening lessons. It has demonstrated vividly that the political process can turn a bold innovation into a mere embellishment of current programs. And it has shown that a domestic free zone can easily become the source of severe competition to neighboring firms and merely reallocate existing national economic activity for the benefit of the target population.

A domestic free zone, or enterprise zone, must therefore be designed to generate new economic activity based predominantly on the indigenous resources of an otherwise stagnant area. Only then can a zone become a national asset. In Britain, there was scarcely an attempt to refine the free zone mechanism to achieve this objective. In America, the odds may still be against the political process bringing forth an enterprise zone of the kind, but the intense debate over the zone proposal is certainly causing those odds to shorten.

NOTES

1. For a history and analysis of the enterprise zone concept in Britain and the United States, see Stuart Butler, *Enterprise Zones: Greenlining the Inner Cities* (Universe Books, New York, 1981).

2. The Fabian Society, established in 1884, became the driving intellectual force behind the creation of the British Labor Party. Its members, who have included George Bernard Shaw, and Sidney and Beatrice Webb, repudiated the Marxist class struggle and argued that socialism could be achieved through gradual, democratic means. Although no longer a key force in the Labor Party, it continues to provide an important intellectual focus for moderate socialists.

3. Even if resources were "imported" into the zone, arguably there would be little cost if the prevailing national regime of tax and regulation would cause them to stay unused if they were outside the zone.

4. See Roger Vaughan, *The Urban Impacts of Federal Policies: Volume II, Economic Development* (Rand Corporation, Santa Monica, California, 1977), Roger Schmenner, *The Manufacturing Location Decision* (MIT, Cambridge, Mass., 1978), and Bennett Harrison and Sandra Kanter, "The Political Economy of States' Job Creation Incentives," *AIP Journal,* October 1978.

REFERENCES

BIRCH, David (1980) Job Generation in Cities. Cambridge, Mass.: MIT.

BIRCH, David (1979) The Job Generation Process. Cambridge, Mass.: MIT.

CALLIES, David and TAMASHIRO, Gail (1983) "Enterprise Zones: The Redevelopment Sweepstakes Begins." The Urban Lawyer (Winter) 15, 1:232.

The Economist (1980) March 29.

HALL, Peter (1977) Speech to the Royal Town Planning Institute, June 15, quoted in BUTLER, Stuart (1981) Enterprise Zones: Greenlining the Inner Cities. New York: Universe Books.

JACOBS, Jane (1969) The Economy of Cities. Random House.

JACOBS, Jane (1961) The Death and Life of Great American Cities. Random House.

Part III

State and Local Approaches

8

Urban Economic Development:
A Zero-Sum Game?

FRANKLIN J. JAMES

☐ INTEREST AMONG federal, state, and local policymakers in economic development has grown greatly over the past decade. Factors accounting for the new interest include the following:

— faltering national economic performance and a renewed appreciation of the importance of economic strength in shaping the overall quality of life;

— the spreading out of economic activity among communities and regions, which has created severe dislocations in some areas but brought new opportunities for economic development in others; and

— aggressive efforts by the federal government to stimulate public private partnerships for economic development; these efforts were initiated during the Carter administration, and persist (though in a different form) under President Reagan.

Paradoxically, many urban economic development programs are in grave difficulty, despite the current interest. Federal government programs are most endangered; all of them have experienced deep budget cutbacks since 1980. The Economic Development Administration has been proposed for elimination by the Reagan administration. More significantly, appropriations for HUD's highly popular and apparently effective Urban Development Action Grant Program have been cut by one-third, and actual spending under the program has fallen even more rapidly. Important economic development appurtenances of HUD's Community Development Block Grant Program (e.g., Section 108 loan guarantees) have been proposed for elimination. Even the Small Business Administration has been cut back.

Some cutbacks in these federal programs were to be expected, given disarray in the federal budget. However, the problems faced by federal economic development programs have a much deeper origin than simple budget stringency. Basically, the programs lack a clearly articulated rationale backed up by hard evidence.

The main objective of existing urban economic development programs is to contribute to greater *equity* in the distribution of economic opportunity among groups of the nation's population. For example, to increase job opportunities for minorities, the poor, and for structurally unemployed persons left behind by change in the economic functions of their communities. Advocates of the programs are unable to argue convincingly that economic development is a cost-effective way to accomplish these goals. The programs have been highly vulnerable to criticisms from both the political right and left that they represent wasteful efforts to prop up the economies of obsolete communities (for example, President's Commission, 1981). *Because* the programs are targeted to economically distressed communities, there is inevitable doubt that the programs strengthen the U.S. economy, and fear that they may weaken the economy by drawing business to inefficient locations.

Historically, geographically targeted urban economic development programs have waxed under Democratic presidents and waned under Republicans. However, it would be wrong for advocates to assume that economic development programs will prosper when Democrats regain the presidency. There is growing interest within the Democratic party in mounting some kind of national industrial policy aimed at undergirding the economic strength of the nation. (Blumenthal, 1983). It is by no means clear that current economic development programs for distressed communities would play an important part in such a policy, unless a convincing case can be made that the programs do foster national productivity and growth, and that they are a cost-effective way to accomplish their redistributive goals. The first section of this chapter will briefly examine and challenge the rationale of current geographically targeted programs. It contends that the theoretical foundations for targeted economic development policies are ambiguous and evidence of their effectiveness is scant. As a result of the weak rationale, urban economic development programs continue to be targets for budgetary cutbacks. The second section describes how local development efforts evolved since 1980 under the impact of national program cutbacks. Finally, the chapter identifies the kinds of evidence that might contribute to a convincing rationale for federally financed economic development programs.[1]

Hopefully, this chapter will energize *advocates* of urban economic development to do the homework that is required to demonstrate the efficacy of the programs. Unless this basic homework is done, program opponents can persist with impunity in cutting back the resources available for investment in economic development efforts.

A NATIONAL URBAN ECONOMIC DEVELOPMENT POLICY?

As suggested above, the most important argument for the programs is that there are greater *needs* for jobs and business investment in some communities than in others; that is, the programs are hoped to redistribute economic activity among population groups and communities. However, the clearest political message of the last few years is that redistributive programs lack political support unless they also enhance overall national economic progress. This current political stance is broadly consistent with historic national policies in the United States regarding economic development.

A HISTORICAL PERSPECTIVE

The history of geographically targeted economic development through 1980 in the United States can be divided into three broad phases. Up to the 1920s, the focus of such economic development activities was on settling the continent and establishing the United States as a major industrial world power. During this period, government economic development efforts provided basic infrastructure needed in *developing* areas of the nation and *developing* sectors of the national economy. Another focus was on educating the American population at large. The perceived payoff to such investments was straightforward: national economic growth and increased economic opportunity for both investment and jobs.

A second major phase extended from about 1930 to 1965 or so. Depression and war strengthened the role of government in shaping national economic progress. However, a new focus of economic development activity emerged: efforts to stimulate the regional economy of the South (including northern portions of Appalachia). These regional economic development efforts enjoyed widespread popular support over a long period of time.[2]

Poverty and the economic backwardness of the South were severe during this second phase, and one goal was to redistribute economic opportunity to people in this depressed region. However, successful development efforts were hoped (accurately, it seems) to expand overall national business investment opportunities and enhance business access to what was clearly an underemployed work force. During this second phase, targeted economic development efforts simultaneously increased the economic opportunities of need people *and* increased the *national* wealth.

The mid 1960s to the present is the period during which most urban economic development programs were created, and comprises a third phase. This third phase differs sharply from earlier periods, principally in terms of an increased focus of programs on accomplishing redistributive goals and a diminished emphasis on fostering overall economic progress.

The geographic focus of economic development in the current phase is blurry, but has generally narrowed to distressed cities, communities, and neighborhoods.[3] The narrowing geographic focus of economic development programs reflects a change in the types of economic distress with which the programs are trying to deal.[4] The current recession as obscured the fact that the most severe problems of structural unemployment, underemployment, and poverty are found among three groups of the population: blacks, Hispanics, and women heading families with children (U.S. Commission, 1982). These are the groups on whom economic development efforts have focused resources in recent years. These groups are increasingly concentrated in big cities where they are isolated from economic opportunity by several factors, including

— discrimination in employment;

— low skills and (in the case of some women) family barriers to work; and

— loss of economic activity in their communities (Bradbury et al., 1982).

In this third phase, targeted economic development activities may have become invidious in their impacts. Geographically targeted economic development activities have been implemented to stimulate employment and investment in many distressed big cities and away from their more affluent suburbs. Within cities, economic development resources have been used to stimulate impoverished neighborhoods rather than economically healthy or vibrant ones. Economic development programs have been implemented explicitly to redistribute wealth to minorities and the poor.

The bottom line is that national economic development programs have emerged during the third phase with the following:

— a relatively narrow constituency of direct beneficiaries;

— a rationale based largely on redistributive goals; and

— a tough policy agenda of ameliorating economic distress among highly troubled segments of the population and of the national geography.

THE STRUGGLE FOR AN ECONOMIC RATIONALE

At present, there are theoretically plausible reasons suggesting that urban economic development programs can contribute to macroeconomic goals of higher output and employment. However, theoretically plausible cases can also be made that targeted development programs simply shift benefits from one person to another or from one area to another. There is no convincing empirical evidence that urban economic development as currently practiced is more than a zero sum game.

Employing the unemployed. The classical justification for geographically targeted economic development programs has been that they assist unemployed or underemployed persons in finding productive work. This was true of regional economic development programs from 1930 to the early 1960s. There are no convincing measures of the employment benefits of current programs for the disadvantaged populations of urban areas. However, their is little doubt that today's programs are less effective in generating new job opportunities than were regional programs of the past.

The lesser effectiveness of today's programs is attributable in part to their finer-grain geographic targeting. Municipal or county boundaries used to define distressed areas in today's programs do not correspond to anyone's concept of a labor market. Metropolitan areas or urbanized areas conform much more closely to economists' notion of a labor market. This basic fact has two obvious implications. First, some government economic development resources will be invested in distressed communities that are located within healthy or even tight metropolitan labor markets. When this occurs, few net new job opportunities are likely to be created unless new jobs attract migrants or long distance commuters from outside the metropolitan area.

Second, economic development projects successfully targeted to distressed cities within troubled labor markets are unlikely to be as productive in generating net new jobs as the projects would be if they were targeted to more competitive locations. Substantial barriers impede successful economic development in the nation's distressed cities, particularly when the city lies within economically depressed metropolitan areas.[5] Such barriers make job creation risky and expensive in public subsidies.

Another, more basic factor limiting the potential employment benefits of today's economic development programs is that structural unemployment problems today are simply much tougher to solve than they were in the past. Job creation is not a sufficient strategy for addressing the employment problems of racial or ethnic minorities, single parents, or the impoverished. Training programs, fair employment programs, and a variety of social services are needed to complement job creation efforts. So far, few communities have had the resources, management skills, or information needed to undertake coordinated programs involving these disparate efforts.

Given the severity of the employment problems facing urban disadvantaged populations (and the targeting problems discussed above), it is at least arguable that many if not most of the jobs created by economic development programs will not materially ameliorate structural unemployment.[6] A recent, careful evaluation of the early experience of the UDAG program reports that only one in four persons hired as a result of the projects were unemployed at the time they took the job, and only one in ten were participating in or eligible for assistance under the CETA program. (U.S. Department of Housing and Urban Development, 1982: 65).

Fiscal benefits. Current targeting practices in urban economic development programs could conceivably be justified by fiscal benefits of projects; that is, it could (and sometimes has been) suggested that current targeting schemes are designed to maximize the *overall fiscal and job benefits* of the programs, not merely their employment benefits alone. Private investment leveraged in distressed cities by the programs adds property tax base and personal income that augments other tax bases. Unfortunately, virtually no evidence exists with which to judge how important such fiscal benefits really are. One study reports that 10 of 11 selected federally assisted economic development projects had favorable fiscal impacts in New Jersey communities; that is, that the projects increased local *revenues* more than they did local service *costs* (Burchell and Listokin, 1981). However, as the authors emphasize, this sample of projects was far too small to justify any conclusions about the overall fiscal benefits of other projects in New Jersey or elsewhere. Because federal economic development assistance is frequently accompanied by a variety of local and state aid, including tax exemptions or abatements, it would be optimistic to presume that the findings of this study are generally applicable.

The "counter-balancing" argument. A third line of argument suggests that targeted economic development efforts can increase the economic efficiency of the U.S. economy by helping to neutralize some of the antiurban impacts of a wide variety of federal programs.

This is a strong argument because correcting distortions due to other policies can contribute to greater economic efficiency. A number of experts have pointed out that federal policies in areas of taxing, housing, highways, and mass transit have inadvertently undercut the economies of big cities. For example, federal tax policies have long fostered business investment in *new* plants and equipment. In the process, the policies have hastened the movement of business from communities unable to compete for such *new* investment to communities attractive to business (Peterson, 1979; Kaplan et al., 1981). Recent research offers some empirical support for this hypothesis (Hulten et al., 1982).

Policies that foster disinvestment in cities and neighborhoods waste physical capital: infrastructure, factories, housing, stores, and so on. Such policies also waste human capital by contributing to unemployment and underemployment. It is thus plausible that economic development programs that help neutralize the antiurban biases would limit such waste and contribute to a more efficient national economy.

However, no analyst has been (or probably will be) able to report how much infrastructure investment has been wasted by programs with an antiurban bias, or how many people face unemployment, underemployment, or dependency because of programs undercutting urban economies. Indeed there is startling lack of hard evidence showing that employment opportunities are truly inferior for disadvantaged persons in distressed cities than elsewhere (James and Blair, 1983).

By and large, advocates of urban economic development have been content merely to point out that economic development programs could help neutralize some of the adverse urban impacts of other programs. This position begs the question of whether urban economic development efforts are the best way in which to do this. A number of analysts argue that old big cities have lost most of their economic functions. Indeed, the pace of job loss in the nation's big, distressed cities has been both rapid and accelerating during the past decade (HUD, 1980). If cities have lost much of their economic functions, why does it make sense to subsidize new long-term business investment in the cities? Perhaps, as many argue, it makes greater sense to offer cities more flexible aid designed to compensate the cities and their residents for adverse impacts and to help them cope with economic change.

The "small business" or infant industry rationale. Small business remains a puissant political force in the United States, and advocates of targeted economic development efforts have attempted to enlist this constituency. In substantive terms, it is argued that

— small business is a major producer of business innovations in the U.S. economy (i.e., new products, new production methods, etc.);

— small business accounts for the bulk of *new* jobs added in the U.S. economy.

— a major underlying failure in distressed city economies is the absence of business innovation; therefore,

— targeted economic development programs that foster investment in small business may stimulate productivity and growth in both distressed cities and in the U.S. economy (Morrison, n.d.; Hanson, 1981; Kieseschnick, 1979).

Advocates of this position frequently denigrate aid to larger enterprises as unneeded, arguing that the enterprises can fend for themselves. Indeed, there is little hard evidence that the location or investment decisions of large enterprise have been much affected by government inducements or incentives (Schmenner, 1978). However, neither is there much support for the positive arguments made for helping small business.

Available evidence contradicts the widely believed assertion that small businesses generate a disproportionate number of new jobs. The best recent evidence suggests that small businesses generate about the same proportion of new jobs in the national economy as they account for among existing jobs (Armington and Odle, 1982). One earlier study reported that small business accounted for the bulk of new jobs in the national economy. However, this early study was based on flawed data and included small branch plants or offices of larger firms as "small business" (Birch, 1978).

The recent research shows that small businesses account for a somewhat larger proportion of the job growth in big cities of the North than elsewhere in the economy. However, this seems attributable to the weakness of demand for labor on the part of larger firms in these cities, not to any special economic potential of small business (Armington and Odle, 1982).

As a practical matter, few federal economic development efforts have much helped small business. When they have, the results have often been bad. Major targeted federal grant or loan programs are seldom efficient in delivering aid to small business, because of the potentially high transaction costs involved. UDAG, for example, is best at providing aid to large investment projects that have impacts commensurate with the public costs of reviewing applications and monitoring implementation.

More generally, government programs for small business (such as the loan guarantee programs of SBA or SBICs/MESBICs) have seldom shown significant capacity for aiding innovative promising small business. Many of the loan guarantee programs have experienced relatively high default rates (Bates, 1980). MESBICs and minority loan guarantee programs have earned a reputation for extremely high failure rates and little even anecdotal reports of outstanding success.

RECENT TRENDS IN STATE AND LOCAL ECONOMIC DEVELOPMENT EFFORTS

While there is little strong evidence that national economic development efforts are appropriate or effective, there are plentiful reasons to believe that federal programs are crucial if urban economic development efforts are to happen. Put another way, existing evidence does not prove that the job of economic development is worth doing, but there is little doubt that the job will not be done effectively without potent national programs.

The experience of the last three years strongly suggests the importance of federal programs. Recent budget cutbacks in federal programs have had the obvious consequence of reducing the overall resources available for development activities at the local level. There is no reason to believe that lost federal resources have been made up from other sources. More importantly, the federal government has in the past been a major force for targeting and for equity in the distribution of economic development resources. Recent experience suggests that the loss of such federal leadership would diminish the targeting of state and local economic development activities.

THE TOOLS AND GOALS OF THE STATES

States have more authority and more impacts on local urban economic development efforts than does the federal government (Vaughan, 1979). States define the legal authorities of localities to undertake various kinds of activities and define the overall fiscal responsibilities and resources of localities. The current recession has increased the willingness of some states to invest money in economic development efforts. It is sometimes suggested that states could (and perhaps should) be called upon to take on the responsibility of urban economic development.

Some states have been active in fostering the economic development of cities. Massachusetts and Michigan are widely noted examples (Pihl, 1979; Pilcher, 1983). However, most states have focused their efforts on stimulating the overall economies of their states rather than those of specific distressed places within the states. There is no doubt that state efforts have become less targeted on distressed cities in recent years. For example, there is a rush among the states to compete for various high-tech industries. North Carolina got into this game years ago with its Research Triangle Park. Michigan is assembling money for a $230 million Center for Robotics Excellence to be housed at the University of Michigan. Arizona is building a Center for Engineering Excellence. Colorado has followed the classical Sunbelt economic development strategy of attracting a multibillion dollar Air Force command center in Colorado Springs for war-related space programs (Speer, 1983). Like other states, Colorado is also beginning to upgrade enginering programs in its state university to attract and service high tech electronics industries.[7] Few (if any) state programs are designed to foster high-tech activities in distressed cities.

Significantly, the most important recent targeted economic development initiatives of the states have been associated with urban enterprise zones. The *national* program proposed by Congressmen Kemp and Garcia and supported by the administration has gone nowhere. The proposal suffers both from legitimate doubts about its likely effectiveness and from the unwillingness of potential constituency groups to support a new untried program when existing effective programs are in budget trouble (Sternlieb, 1981; Howe, 1981; Butler, 1981).

Despite the limited progress of the national legislation, a substantial number (10, as of July 1982) of states have passed enterprise zone laws in anticipation of national action (Gold, 1982). These laws provide for a variety of state tax incentives to foster private business investment in zones. The tax incentives include state income tax credits for hiring or investment, as well as sales tax, property tax, or other kinds of tax breaks (Gold, 1982). More importantly, the state laws generally incorporate geographic targeting provisions similar to those used in existing federal programs (e.g., UDAG) or proposed in national enterprise zone legislation (Gold, 1982). It is highly unlikely that such targeting provisions would have been selected or possible for the states in the absence of federal government leadership.

LOCAL INITIATIVES

The importance of national leadership and resources is also apparent in recent trends in local economic development efforts. At base, today such efforts are strapped for funds and making little headway. There is little doubt that local efforts (like those of the states) are becoming less targeted on both distressed cities or on their disadvantaged residents.

To be sure, a number of economic development initiatives have been recently made by localities, foundations, and the private sector. This activity is exciting evidence of widespread concern with urban economic development. However, much of the effort seems designed to maintain existing economic development capacity in cities and neighborhoods in the hope that real resources for action may be forthcoming in the future.

Charitable investments by the private sector and foundations. Both private business and foundations have made increased investments in urban economic development in recent years. One way the private sector has helped is through modest scale investments in capable development organizations, especially nonprofit or community-based organizations outside city government.

Probably the best known national organization involved in such investments is the Local Initiatives Support Corporation (LISC), which was established by the Ford Foundation with the support of a number of major private companies. LISC raises investment funds from such national sources and distributes the money to cities with the capacity to raise matching funds from local sources. LISC is reported to have accumulated $40 million in funding (Berger, 1982). LISC's goals include

— "helping forge a productive and continuing alliance between community organizations and the local and national private sector."

— "strengthening local capacity [by] assisting organizations to increase and safeguard their assets. The focus is on ventures and projects that produce equity and revenue for the organizations developing them."

— Support is tailored "to encourage and aid local organizations in achieving the business discipline necessary to raise and invest their own funds optimally" (Berger, 1982: 174).

Local LISCs have been formed in Boston, Chicago, Cleveland, and Washington, D.C. Loans and grants from the national LISC to these local LISCs are reported to have been matched $5 to $1 by local resources (Berger, 1982).

Of course, LISCs are not alone. A large number of private businesses, leaders, and local foundations have undertaken "social" investments to foster economic development. For example, James Rouse has formed the Enterprise Foundation, with business and foundation support. This organization operates a for-profit development subsidiary that is hoped to generate ongoing revenues for the foundation (Berger, 1982).

However, the potential of such actions to support economic development on a significant scale is distinctly limited. Their scale in hard economic terms has been modest so far. They are unlikely to grow to major proportions in the future. Activities funded through charitable giving of corporations or individuals are constrained by national tax policy changes made in 1981 that make such giving far more expensive in after-tax terms to donors (Saloman, 1981; Clotfelter and Stenerle, 1981). Moreover, recession and high interest rates have strapped many potential givers to such an extent that their giving will be blighted for years to come. There is a trend among foundations to use their endowments for program-related investment in urban economic development for generating income to fund future grants. This trend may grow in the future but will be limited by the sometimes great risk entailed in urban economic development.

In the present environment, greater investments in urban economic development would have to come at the cost of diminished charitable giving for other purposes. Would (or should?) it win out against other programs designed to improve the quality of life for the disadvantaged?

Mining the tax system. While proposals for urban enterprise zones are by far the best known efforts to utilize the federal tax system to foster investment by private business, they are not the most important.

For the present, national provisions of federal tax law are proving a far more fruitful source of support for local economic development activities. Three provisions in particular are highly valuable:

(1) investment tax credits for the renovation of older industrial and commercial facilities;
(2) tax incentives for the preservation of historic structures, including residential properties; and
(3) Accelerated Cost Recovery System depreciation provisions that have opened the door for a wide variety of property leasing arrangements.

These national tax incentives are being used with increasing imagination by local governments.[9] For instance, Baltimore has used leasing arrangements to finance the city's Culinary Arts Institute, a training facility for restaurant workers. Springfield, Ohio worked out a deal under which regional investors purchased then leased back the local International Harvester plant, in an effort to keep the plant open, even if International Harvester should fail. Transit authorities from a number of major cities have used Safe Harbor Leasing to acquire rolling stock.[10]

Investment tax credits for the rehabilitation of older commercial and industrial properties (and for historic housing) are proving another rich source of resources for economic development. These tax credits range from 15 to 25 percent of rehabilitation investments, and thus offer sizeable lures for private sector investors to involve themselves in improvement efforts in older neighborhoods.

While these incentives provide a great deal of potential resources for use in urban economic development programs, they are equally available for programs outside of distressed cities. They thus offer no substitutes for targeted economic development programs.

Public-private packaging of resources. Public-private partnerships provide a third related trend that has flowered during the 1980s. The term "public-private partnerships" is used here to refer to explicit collaborative efforts between government and business to facilitate major development goals shared by both sectors (CED, 1982).

For example, the Denver Partnership is rapidly becoming a nationally known leader in the innovative packaging of existing public resources and in the formation of effective partnerships between the public and private sectors. The Partnership is a private nonprofit organization whose board of directors includes representatives of virtually every heavyweight business concern in the Denver metro area. As the city cut back its planning and development capacity during the lates 1970s and early 1980s, the Partnership expanded its capabilities in urban design, planning, and implementation. For example, the Partnership has been deeply involved in the development of historic preservation plans in portions of Denver's downtown. It is also played a key role in developing incentive zoning programs to foster retail investments along the downtown's new 16th Street Pedestrian Mall, a $75 million project along Denver's major downtown commercial street. The Partnership is managing and maintaining the Mall, under contract with the city, and worked with the city and downtown property owners to establish a Mall Maintenance District to pay for the operating costs of the Mall.

In these roles, the Partnership is filling in gaps in public capacity, both fiscal and substantive. The Partnership draws on a wide range of revenue sources, including membership dues, foundation grants, government grants, and contracts. Just as importantly, the political and economic clout of is members means that the Partnership can mobilize local and regional capital for major projects.

Of course, there are a number of similar organizations with distinguished track records around the country. Central Atlanta Progress has played a key role in downtown development in that city. Similarly, private business organizations played a major role in developing and implementing redevelopment efforts that have been underway in Baltimore for 30 years.

However, such groups are growing in their power to shape local economic development activities as government fiscal capacity diminishes. Their growing roles are likely to improve some aspects of the quality of local economic development efforts. The substantial private sector role (both financial and leadership) in such partnerships creates the substantial likelihood that projects that result will pass private sector tests of profitability and risk. Another hopeful aspect is that the partnerships stretch the power of the resources of both the public and private sectors to accomplish their shared and separate objectives.

Far from being a substitute for federal programs, however, such partnerships have been facilitated by the past availability of federal aid. Moreover, such partnerships are generally easier to put together in economically healthy environments (such as Denver) where both the public and private sectors can marshall significant resources for development projects.

Perhaps most importantly, public-private partnership efforts cannot be expected to have the same commitment to social equity as endeavors can have when the initiative lies in the public sector and the federal government in particular. In the absence of federal injections of risk capital, such partnerships are likely to focus increasingly traditional types of activities offering few direct benefits to disadvantaged persons.

GOALS FOR FUTURE RESEARCH

Hopefully, the message is clear that proponents of geographically targeted urban economic development programs must support a strong program of basic research if such programs are to outlast the next few years. The research needed would (at a minimum) document the potential economic impacts of the programs, show they can contribute to national economic objectives, identify potential beneficiaries, and demonstrate methods to help make sure that intended beneficiaries actually benefit. Such research could provide a realistic body of expectations among the public and among policymakers for judging whether the programs are doing their jobs or not (HUD, 1982).

THE FUTURE ECONOMIC POTENTIAL OF DISTRESSED CITIES

One obvious goal for research is to deepen our understanding of the real comparative advantages of the economies of distressed cities and how the economies of these cities are likely to evolve in the absence of significant national economic development programs. Such information remains seriously lacking, though a few cities or metropolitan areas have conducted useful analyses of particular local economies (Gurwitz and Kingsley, 1982; Bluestone, 1979). Information describing and measuring the structural transformations going on in the economies of distressed cities could help defend programs from casual charges that they are subsidizing inefficiency. Such information could also provide a baseline against which to measure the potential impacts of development programs (Vidal et al., 1983).

REALISTIC SCENARIOS OF THE IMPACTS OF ECONOMIC DEVELOPMENT PROGRAMS

Thus far, policy analyses of economic development programs have generally been content to establish the importance of a strong economic base to cities (HUD, 1980). Evaluations of programs have focused on measuring the impacts of particular projects subsidized under the programs (HUD, 1982). The obvious failing of such analyses is their inability to specify how much of a dent any feasible economic development program could make in the overall scale and distribution of urban distress. It makes less sense to fight for such programs if at most they can make a tiny difference than if they can have a substantial impact. Moreover, the data and methods required to do such an analysis could conceivably enable an evaluation of alternative economic development strategies.

High powered methodologies such as computer simulation modeling could conceivably facilitate such an analysis, but are risky and not the only alternative.[11] One recent study used relatively straightforward methodologies to simulate the potential impacts of urban programs in Cleveland (Bradbury et al., 1981).

Particular emphasis should be placed in such studies on judging the magnitudes of the secondary impacts of economic development on city and state fiscal well-being and on revitalizing neighborhoods and strengthening the population bases of cities. That such benefits are important has been a major argument for targeted urban economic development.

CONCLUSIONS

None of the foregoing should be taken to mean that the author does not believe that a strong case cannot be made for targeted urban economic development programs. Indeed, the author has advocated such programs in the past, and continues to do so. What the above should be taken to mean is that supporters of the programs have a responsibility to test the programs against the facts and show that they measure up.

REFERENCES

ARMINGTON, C. and M. ODEL (1982) "Small business—how many jobs?" Brookings Review (Winter).

BATES, T. (1980) "Effectiveness of the SBA in financing minority business." Presented at a conference of the American Economic Association, Denver, CO.

BERGER, R. A. (1982) "Conclusions: partnership innovations," in Investing in America: Initiatives for Community and Economic Development. Washington, DC: The President's Task Force on Private Sector Initiatives.

BIRCH, D. L. (1978) "The processes causing economic change in cities." Prepared for a Department of Commerce Roundtable on Business Retention and Expansion, Washington, DC.

BLUESTONE, B., P. JORDON, C. PEPIN, and M. SULLIVAN (1979) The Aircraft and Parts Industry in New England. Cambridge, MA: Joint Center for Urban Studies, Cambridge University.

BLUMENTHAL, S. (1983) "Drafting a democratic industrial plan." New York Times Magazine (August 28).

BRADBURY, K. L., A. DOWNS, and K. A. SMALL (1982a) Urban Decline and the Future of American Cities. Washington, DC: Brookings Institution.

——(1982b) Futures for a Declining City: Simulations for the Cleveland Area. New York: Academic Press.

BUNCE, H.L., R. BENJAMIN, and S. G. NEAL (1983) The Effects of 1980 Census Data on Community Development Funding. Washington, DC: U.S. Department of Housing and Urban Development.

BUNCE, H. L. and S. G. NEAL (1983) "Trends in city conditions during the 1970s." Washington, DC: U.S. Department of Housing and Urban Development.

BURCHELL, R. W. and D. LISTOKIN (1981) "The fiscal impact of economic-development programs: case studies of the local cost-revenue implications of HUD, EDA, and FmHa projects," in G. Sternlieb and D. Listokin (eds.) New Tools for Economic Development: The Enterprise Zone, Development Bank, and RFC. New Brunswick, NJ: Rutgers University Center for Urban Policy Research.

BUTLER, S. (1981) "Enterprise zones: pioneering in the inner city," in G. Sternlieb and D. Listokin (eds.) New Tools for Economic Development: The Enterprise Zone, Development Bank, and RFC. New Brunswick, NJ: Rutgers University Center for Urban Policy Research.

CLOTFELTER, C. T. and C. E. STENERLE (1981) "Charitable contributions," in H. J. Aaron and J. A. Peckman (eds.) How Taxes Affect Economic Behavior. Washington, DC: Brookings Institution.

Committee for Economic Development (1982) Public-Private Partnership: An Opportunity for Urban Communities. Washington, DC: Committee for Economic Development.

EWING, R. (1978) Barriers to urban economic development. Washington, DC: U.S. Congressional Budget Office.

GOLD, S. D. (1982) State Urban Enterprise Zones: A Policy Overview. Denver, CO: National Conference of State Legislatures.

GURWITZ, A.S. and G.T. KINGSLEY (1982). The Cleveland Metropolitan Economy. Santa Monica, CA: Rand Corp.

HANSON, D. (1981). Banking and Small Business. Washington, DC: Council of State Planning Agencis.

HOWE, G. (1981) "Liberating free enterprise. a new experiment," in G. Sternlieb and D. Listokin (eds.) New Tools for Economic Development: The Enterprise Zone, Development Bank, and RFC. New Brunswick, NJ: Rutgers University Center for Urban Policy Research.

HULTEN, C., G. PETERSON, and R. SCHWAB (1982). The Regional and Urban Impacts of Federal Tax Policy. Washington, DC: Urban Institute.

JAMES, F.J. and J. P. BLAIR, (1983) "The role of labor mobility in a national urban policy." Journal of the American Planning Association.

KAPLAN, M., R.S. PHILLIPS, and F.J. JAMES (1982) The Regional and Urban Impacts of the Administration's Budget and Tax Proposals. Washington, DC: Joint Economic Committee of the U.S. Congress.

KIESCHNICK, M. (1979) Venture Capital and Urban Development. Washington, DC: Council of State Planning Agencies.

MILKMAN, R.H. (1972) Alleviating Economic Distress: Evaluating a Federal Effort. Lexington, MA: D.C. Heath.

MORRISON, J. (n.d.) Small Business: New Directions for the 1980s. Washington, DC: National Center for Economic Alternatives.

National Council for Urban Economic Development (1982) "Leasing as a tax incentive tool," CUED Legislative Report (July 28).

——(1982) "Packaging the rehabilitation tax credit in an economic development program." CUED Legislative Report (March 3).

PETERSON, G.E. (1979) "Federal tax policy and urban development," in B. Chinitz (ed.) Central City Economic Development. Cambridge, MA: Abt Books.

PIHL, M.H. (1979) "Urban economic development strategies: the state role in Michigan," in B. Chinitz (ed.) Central City Economic Development. Cambridge, MA: Abt Books.

PILCHER, D. (1983) "Michigan: the road to recovery," in National Conference of State Legislatures, State Legislatures. January.

President's Commission for a National Agenda for the Eighties (1981) Urban America in the Eighties. Washington, DC: Government Printing Office.

SALAMAN, L. (1981) Federal taxes and charitable giving. Washington, DC: Urban Institute.

SCHMENNER, R. (1978) The Manufacturing Location Decision: Evidence from Cincinnati and new England. Cambridge, MA: Harvard-MIT Joint Center for Urban Studies.

SPEER, R.D. (1983) "High hopes for high technology" National Conference of State Legislatures, State Legislatures. January.

STERNLIEB, G. (1981) "Kemp-Garcia Act: an initial evaluation," in G. Sternlieb and D. Listokin (eds.) New Tools for Economic Development: The Enterprise Zone, Development Bank, and RFC. New Brunswick, NJ: Rutgers University Center for Urban Policy Research.

U.S. Commission on Civil Rights (1982) Unemployment and Underemployment Among Blacks, Hispanics, and Women. (Washington, DC: Government Printing Office.

U.S. Department of Housing and Urban Development (1980) The President's National Urban Policy Report: 1980. Washington, DC: Government Printing Office.

——(1982) An Impact Evaluation of the Action Grant Program. Washington, DC: Office of Policy Development and Research.

VIDAL, A.C., R.S. PHILLIPS, and H.S. BROWN (1983) The Growth and Restructuring of Metropolitan Economies: Decentralization and Industrial Change During the 1970s. Cambridge MA: Harvard MIT Joint Center for Urban Studies.

9

Competition and Cooperation among Localities

WILLIAM B. NEENAN
MARCUS E. ETHRIDGE

☐ IN EVALUATING the prospects for economic development, it is necessary to address the issue of interlocal competition and coordination. State and local governments as well as the individuals residing within these jurisdictions both compete against as well as cooperate with one another. In this chapter, the question of both intergovernmental as well as intragovernmental competition and cooperation will be discussed and implications of this behavior for both economic efficiency and equity explored. Consequences of competitive as well as of cooperative behavior among local jurisdictions will be viewed as a complex problem in which certain individuals may gain from the same actions that make other individuals lose. Clearly, competitive and cooperative behavior can affect public employees, taxpayers, service recipients, and businesses in significantly different ways.

The interests of these various groups sometimes converge while in other instances they conflict. To avoid semantic confusion, therefore, it is important that we carefully identify which "local" groups we have in mind rather than simply referring to "local government." For example, when we read "Springfield, Ohio outbids Fort Wayne, Indiana" for an International Harvester plant and we conclude that "Springfield gains," do we mean that all of Springfield's residents gain and gain equally? The fiscal condition of the city of Springfield may be improved, individual public sector employees benefited, and the profit position of certain business interests within the city enhanced, but it is possible that fiscal concessions granted to International Harvester may impose a burden on residents of Springfield who are employed neither by International Harvester nor by the City of Springfield. One may ask further whether concessions granted by the

City of Springfield were necessary. After all, Springfield's plant was four decades newer and more efficient than the International Harvester installation in Fort Wayne. Thus, International Harvester rather than Springfield may be the principal "gainer" in this episode.

In short, the promotion of economic development in a given area raises complex questions. It is possible that the competition pressures among states and muncipalities can be destructive and that the benefits and costs of economic development may accrue unevenly to different groups of citizens. In order to have a thorough understanding of the problems created by the growing development pressures it is necessary to address the general issue of competition and cooperation among localities.

The topics to be addressed in this chapter are (1) The Tiebout model and maximizing behavior as a source of competition; (2) concept of "voice" and the value individuals place on a responsive political system; (3) welfare costs imposed on individuals because "voice" is restricted through political consolidation; (4) potential for competition implicit in the nature of the political decision-making process and the production of local government services; (5) suburban-central city intergovernmental competition; (6) tax-base sharings as an example of intergovernmental cooperation; and (7) revenue sharing and intergovernmental competition at high levels of government.

THE TIEBOUT MODEL

The Tiebout model of local public finance can usefully serve to introduce the discussion of local competitive behavior. Tiebout's model consists of seven assumptions from which two conclusions are drawn concerning efficiency properties of the mix and level of local public output. It is assumed that consumer/voters (1) are fully mobile regarding residential location, (2) have complete knowledge of local revenue and expenditure patterns, (3) are able to choose from a large number of communities, and (4) are aware that their location decisions are not constrained by employment opportunities. Further, (5) public goods and services are assumed to exhibit no externalities between communities, and (6) each community is of optimum size with local output produced at the minimum point of a U-shaped average cost function and duplicate communities cloned as required. Finally, (7) the optimum size of communities is maintained by the in- and outmigration of citizens rather than by local communities adapting public sector output to the interest of current residents. In other words, dissatisfied citizens "exit" from communities rather than stay and exercise their "voice" to alter matters.

From these seven assumptions, Tiebout concludes that individuals choose to live in a particular jurisdiction because of the expenditure-tax package available to its residents. By thus sorting themselves out, citizens maximize their own individual welfare and public services are efficiently provided across local jurisdictions. Moreover, the Tiebout Model implies that when economic development occurs, business locations will be made in the same efficient manner. New plants will be built in the municipalities most appropriate for the service needs of each business.

A comparative fiscal advantage, however, cannot be maintained by one jurisdiction over others since the presence of such an advantage attracts new residents with attendant congestion costs. The "Invisible Hand" thus so manages matters in the Tiebout world that explicitly competitive behavior between and within local communities is eliminated. By relaxing a few of the strong assumptions of the model, however, we can allow for some of the practices employed both by political jurisdictions among themselves as well as by citizens within a jurisdiction.

By relaxing the first four Tiebout assumptions, we introduce both the possibility of constraints limiting residential choice and the presence of imperfect knowledge of options available. In other words, we allow into the Tiebout world such phenomena as restrictive fiscal zoning, limited employment opportunities, and the economic and psychic costs of moving from one community to another. In this context, it is possible for a jurisdiction to generate and maintain a positive fiscal advantage by attracting high-income residents as well as business enterprises with high property values while excluding residents and businesses that generate high net public sector costs.

The annexation of a contiguous jurisdiction can generate a net fiscal effect similar to that resulting from restrictive zoning. In the former instance, a fiscally attractive area is incorporated into a jurisdiction, while in the latter, a fiscally burdensome group is excluded, with a positive result for residents of the jurisdiction in question.

An older central city, for example, is likely to gain by annexing a surrounding suburban area whose residents have substantially higher incomes than found in the central city. Muller and Dawson (1976: 67) conclude from their analysis of the expenditure/revenue flow for one year that the city government of Richmond, Virginia improved its fiscal situation by annexing 23 square miles of neighboring Chesterfield County in 1971. Estimated revenue, generated primarily from the property tax, in the annexed area was greater than the costs of providing public services to this area. They attributed this positive fiscal impact primarily to the relative underutilization of public education by the residents in the annexed area. Judging from the fact that annexation referenda are often rejected, we may infer that, contrary to the Tiebout model, these instances are zero-sum interactions in which there are winners and losers.

The results of recent annexation referenda are certainly consistent with this expectation. Eighteen of nineteen annexation plans were rejected during a recent eight-year period by Michigan voters, and Oregon voters rejected all of the annexation plans submitted to them during the mid-1970s (Advisory Commission on Intergovernmental Relations, 1982: 378). Most other states requiring that annexation be approved by referendum have had similar experiences.

Consequently, the Tiebout model may not accurately describe the realities of interlocal competition, and economic development may not automatically occur in ways that optimize the allocation of resources. Tiebout's formulation implies that development projects will be accepted or rejected in accordance with the costs and benefits they entail such that all municipalities accommodate just as much development as is consistent with their citizen's preferences. If restrictive zoning and other factors deriving from the fiscal disparities among communities are significant, equity considerations may require that economic development be partially managed by some central authority.

POLITICAL CONSOLIDATION AND WELFARE MAXIMIZATION

The political consolidation of a metropolitan area, by reducing the menu of tax-expenditure packages available to citizens and business enterprises, is likely to impose a welfare cost on all citizens. The one common level of services provided by the consolidated jurisdiction means that packages cannot be tailored to the various tastes and incomes of the citizens as closely as assumed in the Tiebout model. Bradford and Oates, using 1960 data for 53 communities, have estimated such a deadweight welfare loss from constraining expenditures in these communities to their mean value. This constraint would have resulted in expenditure increases ranging up to 40 percent for some communities and decreases of 25 percent for others. They estimate that each dollar transferred from the budget of the higher spending communities was evaluated at $1.15 and each dollar added to expenditures by the lower-spending communities was evaluated at $0.65. Thus the mean deadweight loss for the contributing community was

$0.15 with the loss $0.35 for the recipient community and a total loss, therefore, of approximately $0.50 for each dollar transferred (Bradford and Oates, 1974: 43-94).

These estimates of the efficiency loss from the consolidation of jurisdictions are quite significant and should warn us that the welfare loss from the reduction in choice is potentially important. However, it should be noted that these estimates are based on the strong assumption that the levels of output in the local jurisdictions are efficient prior to consolidation. Barlow concludes in an analysis based on the median voter model that local school expenditures financed by the property tax are typically inefficiently undersupplied (Barlow, 1970: 1028-1040). In a more recent study, Greene and Parliament (1980) estimate the efficiency loss from the hypothetical consolidation of twelve school districts in Broome County, New York. Rather than assuming that each district was providing an efficient level of education prior to consolidation, they derived estimates of the demand for education based on data from 678 school districts in New York State. Their findings imply substantial inefficiency costs from consolidation but less than would occur if each district had been providing efficient levels that had fully respected citizen demand (1980: 209-217).

QUALITATIVE CONSIDERATIONS

The Tiebout model implicitly adopts "exit" as the equilibrating mechanism in determining the level and mix of public output. As Hirschman has pointed out:

> This is the sort of mechanism economics thrives on. It is neat—one either exists or one does not; it is impersonal—any face-to-face confrontation between customer and firm with its imponderable and unpredictable elements is avoided and success and failure of the organization are communicated to it by a set of statistics; and it is indirect—any recovery on the part of the declining firm comes by the courtesy of Invisible Hand, as an unintended by-product of the customer's decision to shift [1970: 16].

However, we know that citizens do not always "exit"—they often prefer to fight rather than switch. Citizens, furthermore, desire more from their communities than merely a high level of expenditures relative to taxes. They also seek such goals as (1) a responsive political system that can effectively represent the views of constituencies as well as promote a common good, (2) respecting citizen preferences concerning both the mix and level of public services, and (3) high-quality public services.

Williams (1971) has developed a conceptual approach for understanding the exit/voice decision faced by individual residents. As he sees it, the costs of both political participation and relocation rise with the extent of change desired, but political participation is usually less costly than relocation where small changes are involved. Relocation is normally considered only when major changes are at stake. Theoretically, a more efficient political process (often thought to be achieved in smaller municipalities) may make political action more likely than relocation as a response to a wider range of desired changes in policy. In other words, a system of many small municipalities could experience more "voice" and less "exit" than the Tiebout model would suggest.

Again, the complexities involved in citizen decisions regarding location make economic development projects politically conflictual. If the violation of Tiebout assumptions places communities in competitive positions, and if individuals are led to act to influence government decisions "voice," the potential for considerable divisiveness, must be considered. The fact that economic development affects taxpayers, employees, and other segments of developing communities differently suggests the inevitability of this conflict.

POLITICAL SYSTEM RESPONSIVENESS

Public choice theorists consider local governments to be organizations that determine citizen demand for public services, assess payments for them, and provide services. In this perspective, citizens are viewed as served not by "government" but by various public service industries, for example, police and fire protection, education, and water supply. The activities of these various "industries" must be coordinated both as departments within one jurisdiction and as parts of a larger politically fragmented metropolitan system of governments.

Some governments appear to "run smoothly" merely because one dominant group determines policy even though there may remain considerable smoldering citizen dissatisfaction. Other governments appear to be ineffective and fraught with dissension, even though all governments viewpoints have an airing, simply because effective actions are not taken. A truly effective government is one that can articulate accurately the views of diverse constituencies and resolve effectively conflicting viewpoints. Successful conflict resolution can take many forms, all of which assume a modicum of good will and trust. A highly fragmented political system without some overlapping of jurisdictions is extremely vulnerable to institutional failure.

An important institutional development in the past decade has been the emergence of regional councils of governments (COGs) composed of the county and municipal governments of a region. In the 1980s, the number of COGs had increased to over 700, more than double the number of fifteen years earlier. They provide regional planning services and often are the body to approve a wide variety of federal grants. They also offer an arena where areawide problems can be addressed while at the same time respecting the authority of local governments.

QUALITY OF PUBLIC SERVICE

The Tiebout model is concerned primarily with the process through which citizens (and businesses) sort themselves out among local jurisdictions. In the voluminous discussion and testing of the model, scant attention has been given to the dynamics of how the level and mix of expenditures and taxes are actually determined within any jurisdiction—what is the nature of the production function of public services and how does the political process within a jurisdiction respond to diverse constituencies? The model does indeed recognize congestion costs in the provision of public services as implied in the assumption of a U-shaped average cost function with respect to community size. Most of the subsequent analysis based on the Tiebout model has assumed, however, that production functions are identical across jurisdictions and that public sector output is defined simply in terms of per capita expenditures.

If explicit attention is directed to the nature of local government production functions, two problematic characteristics of public output emerge: (1) the quality of a public service is dependent upon the nature of the local population and (2) multiple outcomes rather than a unique one are often associated with any given service.

POPULATION CHARACTERISTICS

Oates has argued that

> for certain key local services such as education, public safety, and environmental quality, the production function is determined by not only the usual direct inputs of labor and capital, but also the characteristics of the individuals who comprise the community. For public safety, for example, a given input of police services will be associated with a higher degree of safety on the streets the less prone are the members of the community to engage in crime. Likewise, the more able and highly motivated are the pupils in a certain school, the higher will be levels of achievement [1981: 95].

To the extent that qualities of the local population indeed have an impact on the quality of public services, exclusionary zoning can be used not merely as a means to increase the fiscal comparative advantage of a community but also "for controlling the composition of the local population so as to enhance the quality of local services" (Oates, 1981: 96). The use of zoning in this capacity obviously has important moral and legal implications and may explain some of the friction observed between blue-collar central cities and their suburbs and the more affluent surrounding ring of communities found in many metropolitan areas.

MULTIPLE OUTCOMES

A potential source of conflict within a community arises from the fact that a service often does not have merely one simple outcome. Take education as an example. Students, parents, and the general public are not interested primarily in the number of "pupil lessons" provided by the public school system. They are looking instead for some outcome we might term "achievement." The question is further complicated when there is not merely one outcome, for example, "achievement," but multiple outcomes that are desired. Some expect public education to provide vocational training, others functional literacy, and still others college preparation. And along with these outcomes schools are regarded as daycare centers and agencies for promoting student promptness, social uniformity, and tolerance. In fact, the disagreement over which of these goals have most been promoted by public education that provides the fire for much of current controversy surrounding public education in this country.

SUBURBAN-CENTRAL CITY
INTERGOVERNMENTAL COMPETITION

A recurring source of conflict within large urban communities is the perception that suburban residents, particularly those that commute downtown to their jobs, exploit the resources of central city. The general argument holds that, among other things, suburban residents use central city streets, police protection, parks, and public services without bearing a proportionate share of the cost of those services. Relatedly, it is also argued that the responsibility for welfare programs extends to all residents of the whole urban area, even if the direct recipients of welfare generally live in the central city. The continuing increase in the costs of many public services has made this issue extremely contentious.

Several approaches to empirical investigation of the issue reveal that there are no obvious answers, and that a finding of exploitation is heavily dependent on certain important assumptions regarding the provision of public services. The earliest research has been indirect correlation analysis designed to measure the relationship between the percentage of the SMSA population living outside the central city and per capita public expenditures in the entire city. Several studies reported positive correlations, indicating that municipal services are a great per capita burden when suburbs are large (Margolis, 1966; Brazer, 1959). Whether this is evidence of exploitation is unclear, however. Most obviously, the central city has a larger commercial and industrial tax base; the relatively high per capita spending may be offset by the added tax base, some of which is arguably made more valuable by virtue of the productivity of suburban commuters.

The ambiguity of this kind of finding has led to more specific inquiries focusing on commuters' impact on the central city. In a study of New York City, Book (1970) compared the value of the benefits received by commuters (the estimated value of the services they collectively consumed plus the marginal cost to the city of supplying those services) and the city's nonresident earnings tax. For 1968, he found that New York's commuters were only paying half of what they should have been paying, thus supporting the exploitation thesis. However, this analysis does not take into account the city sales taxes paid by commuters, the business property taxes paid indirectly by them when they made purchases, or the economic gains generated by their work. Vincent (1971) directly addresses these last points, concluding that commuters create more "gains" than "costs" to the central city. Smith (1972) reaches similar conclusions in a somewhat different study of San Francisco.

Of course, the exploitation issue cannot be resolved through analysis of suburban commuters exclusively since suburbanites that do not commute to the central city also benefit from urban services. In a study of the Twin Cities, Banovetz (1965) addressed this issue by comparing the public service benefits accruing to residents of central city and suburbanite residents to the tax liabilities of each area. He concluded that there was no evidence for the conclusion that either the core cities or the suburbs subsidized each other significantly. However, two important aspects of his research design may have led him to underestimate the net public service gains accrued to suburbanites. Banovetz did not include benefits from highway construction and maintenance, police and fire protection, libraries, and other services that undoubtedly benefit the nonresident population. Moreover, he deemed the benefits of poverty programs to accrue only to direct program recipients. Since Minneapolis and St. Paul have most of the welfare recipients, this assignment of poverty program benefits led Banovetz to "justify" a higher tax burden for the two central cities.

A quite different approach was employed by Neenan (1972). In a study of Detroit and six representative suburbs, he estimated benefit and tax flows for several varieties of public services. Benefits were first valued and allocated on the basis of cost, as in previous studies; however, he adjusts this benefit incidence on the basis of a "willingness-to-pay" index designed to take into account the varying utility derived from public services by citizens in different income levels. (Obviously, wealthier suburban residents are therefore deemed to receive higher utility from most services.) The tax incidence calculation is modified by including the federal offset, tax exporting, and state revenue sharing factors. He concludes that, in the six suburban communities, the average family of four receives a net welfare gain from Detroit of between $4.00 and $22.00 annually.

Much of the criticism of this study focuses upon its use of individual utility evaluation of public services. Ramsey (1972) argues that the exploitation issue is answered with much less ambiguity by determining whether the following calculation yields a negative sum:

city revenue collections from suburban residents
less
cost of city services consumed by suburban residents
plus
cost of suburban services consumed by city residents
less
suburban revenue collections from city residents.

The choice of this analytic strategy over one that attempts to estimate individual utilities derived from municipal services depends on conceptual questions to which there are no obvious answers. Unfortunately, conclusions regarding exploitation are highly sensitive to these research design issues. Moreover, if the individual utility evaluation approach is used, there are several conflicting methods of assigning valuation.

The conceptual subtleties involved in the analysis of suburban-central city exploitation do not prevent the issue from becoming highly politicized, especially as budget constraints are felt more keenly by municipal governments. In a recent newspaper essay, a suburban Massachusetts state legislator described the ethical dilemma he encountered when voting on a bill to raise fees for downtown Boston parking facilities (Barrett, 1983). Over 70% of his constituents contended that they felt "no responsibility" for Boston's financial plight, although another survey discovered 128 instances in 125 randomly selected households of a constituent going to Boston for hospital care, employment, or education. The fact that many of the city facilities charging user fees to suburban residents are exempt from the city's property tax suggests that "out-of-towners" may indeed be able to enjoy at least some benefits without paying a proportionate share of their cost. Whether these imbalances are offset by other consideration (e.g., the suburbanites' contribution to the productivity of businesses in the city) is open to question, but such possibilities can scarcely be expected to defuse the political conflict.

The possibility of suburban exploitation of central city resources has direct bearing on economic development. Businesses that do not have to be located downtown may be led to locate in suburban areas when taxes there are low relative to services received. This situation could have two negative impacts. First, the overall productivity of the area could be reduced since additional commuting and other costs would be incurred whenever these costs are offset by the fiscal disparities between the central city and the suburbs. Second, a major advantage of economic development, increased employment of central city residents, would be less fully obtained when businesses have a special incentive to locate in suburbs. These and other concerns gave rise to the system discussed in the next section.

TAX BASE SHARING

Reschovsky (1980) describes tax base sharing as a mechanism for sharing the fiscal benefits of metropolitan-area growth without disturbing the autonomy of local municipalities. The rationale for tax base sharing is tied to an explicit rejection of the Tiebout model described above. In particular, proponents of tax base sharing contend that mobility restrictions, especially zoning laws, make it highly unlikely that municipal governments will be subject to the clear market forces needed to create an efficient distribution of resources. Moreover, inefficiencies are said to be created when firms are led to make decisions regarding location on the basis of the "fiscal disparities" across municipal jurisdictions.

Reschovsky discusses the issue conceptually and then evaluates the mechanism as it has been implemental in the Twin Cities. Tax base sharing, according to Reschovsky, may arguably reduce horizontal inequities and improve locational efficiencies by tending to equalize the "tax price" for public services. The uneven distribution of wealth leads to means that the tax rate paid by a business or individual tax payer is not simply a function of the costs of providing the service. The "tax price" (the price for unit of service) varies by municipality according to the "wealth of one's neighbors" and the relative concentration of industry and commercial establishments. The resulting inequities are made quite severe as central cities lose their tax base to suburbs while developing a higher concentration of groups and individuals needing more public services. "Tax prices" in the central city rise faster than suburban "tax prices" and businesses make locational decisions based on this factor rather than on the basis of construction costs, commuting expenses, or other factors related to productivity (Reschovsky, 1980).

Obviously, a full metropolitan consolidation with areawide tax rates and centrally administered services would remove the fiscal disparities, but, as Reschovsky points out, efficiency losses would be created by the uniformity of public service levels. Moreover, political opposition to consolidation appears to be insurmountable in virtually every area. Tax base sharing is designed to address the fiscal disparity issue and its associated effects within the structure of a multigovernmental metropolitan area.

The Twin Cities plan was instituted in 1974. The general approach is quite simple. Each local government calculates the *differences* in the assessed value of all the commercial-industrial property within its jurisdiction for the current year and the 1971 base year. Forty percent of the growth (where growth occurred) is contributed to an areawide base. All governments, whether or not they have any growth to contribute, share in the distribution. The formula by which the shares are calculated is positively tied to population and inversely tied to per capita market value of the municipality's taxable property. (Shares are largest for municipalities with large population and small fiscal capacities.)

Under the tax base sharing system, each municipality's tax base is thus its residential and nonresidential property, less its contributed share, plus the share distributed to it. Municipal mill rates are determined by the County Auditor on the basis of each municipality's revenue needs and adjusted tax base. An areawide mill rate is applied to the areawide (jointly contributed) base. Consequently, all commercial-industrial parcels are taxed as two rates: the areawide rate applied to the proportion of the assessed value contributed to the areawide base, and the local rate applied to the remaining value (Reschovsky, 1980).

On the basis of the Twin Cities' experience, Reschovsky concludes that tax base sharing has removed some fiscal disparities, but that it is unclear whether or not locational efficiencies have been improved. Clearly, whatever effects the program could have are dependent on the existence of significant areawide growth. Long-term studies of the system may shed light on tax base sharing as a possible improvement in other areas.

Fox (1981) has provided a useful criticism of the mechanism, however, which suggests that its application could lead to important problems. The essence of his argument is that businesses and industrial operations create negative externalities for their host municipalities, externalities that are accepted because of the tax base expansion they involve. In fact, Fox points out that with a large number of independently operating communities (and under the Tiebout assumption), communities will adjust their tax rates and service packages such that firms will "provide fiscal compensation exactly equal to the value of the negative externalities" they create.

Tax base sharing, according to Fox, reduces the ability of each community to obtain "compensation" for the negative externalities associated with business and industrial firms. Thus, one would expect that many communities would attempt to block new industrial developments within their borders. Some rural communities did begin rejecting industrial projects soon after the plan was implemented in the Twin Cities (Fox, 1981: 276). To the extent that this effect obtains, tax base sharing may disturb the balance of negative externalities and compensation at the local level.

COMPETITION AT HIGHER LEVELS OF GOVERNMENT

The issue of competition among states for industrial expansion projects reveals that states are not homogeneous "winners" or "losers," but that, as indicated at the beginning of this chapter, economic development involves some of both in every jurisdiction. Of course, this complexity is often obscured by the ostensibly common interest that all citizens have in bringing industry to their state, an interest expressed energetically in expensive advertising campaigns by several states (Rosenberg, 1983). The primary instrument manipulated in the heated interstate competition is the business taxation level, and it is increasingly doubtful that tax breaks for businesses have the normally assumed effects on business location.

A 1981 report by the Council of State Planning Agencies concluded that business taxes are rarely important in locational decisions, ranking well below customer access, transportation, labor force skills, and so on (Rosenberg, 1983: 20-21). Amazingly, 80 percent of the businesses responding to the council's study said that they were not aware of the tax abatements they received until *after* they had decided to locate in their current states. These findings are consistent with the fact that low tax rates and economic growth are not consistently related. Rosenberg points out that Massachusetts and California have both high taxes and rapid growth. Conversely, Ohio and Indiana have serious economic difficulties and slow growth, despite their very low tax rates (p. 21). In short, state business taxes are normally not a major factor in locational decisions.

Moreover, there is reason to believe that in the relatively rare cases in which business taxes may make the difference, it is likely to be *capital-intensive* rather than *labor-intensive* businesses that are lured by them. Rosenberg explains that the most common tax breaks reduce property and capital investment taxes while only 18 states offer payroll tax reductions (p. 20). Hence, new business development *attracted by tax manipulations* are not likely to benefit directly, the unemployed citizens of a state as much as businesses attracted for other reasons.

If needed services are slashed as a consequence of tax incentives for industry or if the gap is closed by increasing other taxes, it is quite possible that the majority of a state's citizens will be negatively affected by a "successful" program of business tax incentives. At the very least, it is essential to view such programs critically; specific tax costs must be evaluated with respect to directly attributable development benefits.

CONCLUSION

Whether spontaneous or managed, economic development is usually evaluated with regard to equity and efficiency considerations. The Tiebout world of fully mobile consumers/voters choosing to live in the community that best suites their preference for local revenue and expenditure patterns, is an ideal setting for economic development. Since it posits an optimum allocation of resources, tax rates, and services, any adjustments made as a result of economic development are absorbed efficiently and equitably into the economic system. As mentioned above, there are therefore no persistent fiscal disparities or pressures to compete among local municipalities.

Since there are good reasons to assume that the Tiebout model does not accurately describe reality, political returns that modify local autonomy are often advocated, including metropolitan consolidation and a variety of less comprehensive centralizing changes (e.g., tax base sharing). These reforms vary considerably in their efforts, but, generally speaking, they achieve some improvements in equity at the expense of efficiency. Divergent demand curves are not easily satisfied under consolidated governments (since service delivery is homogenized), and, under a system of tax base sharing, local governments may not be able to be fully compensated for the negative externalities associated with economic development. Economic efficiency is thus reduced. Nonetheless, extreme equity problems may make these or other centralizing reforms viable in some major metropolitan areas.

The larger question of competition among states for economic development raises important questions regarding the distribution of costs and benefits associated with economic development. Olson (1982) provides a useful theoretical framework for explaining the tendency for tax incentives to favor business groups at the expense of the average taxpayer; in simple terms, the diffuse citizenry is unorganized and thus politically weak relative to the groups lobbying for tax benefits as an inducement to locate businesses in a state.

In short, although there is much conceptual and empirical ambiguity concerning the actual fiscal relations among local municipalities and among states, it is clear that the nature of interlocal competition has important implications for economic development. Assessments of the overall value of economic development must take into account the nature and efforts of the interlocal relationships in each metropolitan area.

REFERENCES

Advisory Commission in Intergovernmental Relations (1982) State and Local Roles in the Federal System. Washington, DC: Government Printing Office.

BANOVETZ, J. M. (1965) Government Cost Burdens and Service Benefits in the Twin Cities Metropolitan Area. Minneapolis Public Administration Center, University of Minnesota.

BARLOW, R. (1970) "Efficiency aspects of local school finance." Journal of Political Economy 78: 1028-40.

BARRETT, M. J. (1983) "The out-of-towners." Boston Globe Magazine (August 7).

BOOK, S. H. (1970) "Costs of commuters to the central city as a basis for community taxation." Ph. D. dissertation, Columbia University.

BRADFORD, D. F. and W. E. OATES (1974) "Suburban exploitation of central cities," in H. M. Hochman and G. Peterson (eds.) Redistribution Through Public Choice, New York: Columbia University Press.

BRAZER, H. E. (1959) City Expenditures in the United States. New York: National Bureau of Economic Research.

FOX, W. F. (1981) "An evaluation of metropolitan area tax base sharing: a comment." National Tax Journal 34: 275-280.

GREENE, T. F. and T. J. PARLIAMENT (1980) "Political externalities, efficiencies, and the welfare losses from consolidation." National Tax Journal 33: 209-217.

HIRSHMAN, A. O. (1970) Exit, Voice and Loyalty. Cambridge, MA: Harvard University Press.

MARGOLIS, J. (1966) "Metropolitan finance problems: territories, functions, and growth," in J. M. Buchanan (ed.) Public Finances: Needs, Sources, and Utilization, Princeton: Princeton University Press.

MULLER, T. and G. DAWSON (1976) The Economic Effects of Annexation: A Second Case Study in Richmond, Virginia, Washington, DC: Urban Institute.

NEENAN, W. B. (1972) Political Economy of Urban Areas. Chicago: Markham.

OATES, W. E. (1981) "On local finance and the Tiebout model." American Economic Review 71: 93-98.

OLSON, M. (1982) The Rise and Decline of Nations. New Haven, CT: Yale University Press.

RAMSEY, D. D. (1972) "Suburban-central city exploitation thesis: comment." National Tax Journal 25: 599-604.

RESCHOVSKY, A. (1980) "An evaluation of metropolitan area tax base sharing." National Tax Journal 33: 55-66.

ROSENBERG, T. (1983) "States at war." The New Republic (October 3).

SMITH, R. F. (1972) "Are nonresidents contributing their share to core city revenues?" Land Economics 48: 240-247.

TIEBOUT, C. M. (1956) "A pure theory of local expenditure." Journal of Political Economy G4: 416-24.

VINCENT, P. E. (1971) "The fiscal impact of commuters," in W. Z. Hirsh, P. E. Vincent, H. S. Terrell, D. S. Shoup, and A. Rosett, Fiscal Pressures on the Central City. New York: Praeger.

WILLIAMS, O. P. (1971) Metropolitan Political Analysis: A Social Access Approach. New York: Free Press.

10

Pension Funds and
Economic Development

LARRY LITVAK

☐ PUBLIC AND PRIVATE pension funds should be viewed as attractive sources for certain economic development capital needs. The total assets of pension funds now exceed $1 trillion. These investors have financial characteristics that place them in a much better position to make targeted development investments than almost any other private financial institution. Moreover, the sponsors, trustees, and participants of several funds have expressed a strong interest in making sound investments that would also contribute to their local economies. A recent survey covering 60 percent of public pension funds found $2.3 billion now being targeted to development-oriented investments (Municipal Finance Officers, 1983).

However, there is a legitimate concern that development investing not supplant the primary pension fund objective of providing retirement income. Moreover, the style of investing employed by virtually all pension funds differs greatly from the style that is required for business development investing, and there has been a lengthy history of artificial constraints on pension fund investment practices from which they are only now emerging. To interest and assist pension funds in making development investments, states and cities should develop investment vehicles that offer appropriate management, diversification, controlled risk exposure, and political insulation.

Although there are increasing proposals to use pension fund capital to influence the pattern of private investment, to stimulate job creation and housing construction, and to revive flagging local economies, the record of real effort is still short. Moreover, some of what has happened is not particularly encouraging. The vast majority of pension fund investments made with the express purpose of stimulating new investment in a certain geographic area or form of activity have been so unimaginative that they have merely displaced existing investors, producing no real change in investment patterns. A small number of pension investors have gone to the opposite extreme, making such investments as below market rate mortgage loans, which offer lower rates of return than prudent financial standards would normally allow.

Fortunately, there is an alternative—a way for pension funds to achieve a deliberate development impact without displacing private investors or making financial concessions. This alternative path focuses on making investments in industrial sectors and enterprises that have been "underfinanced" due to gaps and inefficiencies in our financial system—even though they offer rates of return that equal or better prevailing standards (Litvak, 1981).

Capital gaps occur because our financial markets sometimes fail in their basic job of getting capital to projects that offer competitive rates of return relative to the risks they pose. Examples of areas that are now underfinanced include equity financing for emerging companies and long-term debt for modernization and expansion in more mature industries. Each of these areas suffers from a capital gap. The result is a lower rate of economic growth, fewer jobs, and less than desirable housing conditions. Yet, crucially, filling these gaps will not require financial concessions; in fact, these areas can offer investment opportunities more attractive than Fortune 500 corporations.

From a financial standpoint, pension funds are the singlemost appropriate source of capital for many of the nation's most urgent development needs. Many commentators have pointed out the large aggregate size of pension funds, as well as the degree to which they are subject to a measure of popular control. But of equal importance to those concerned with America's economic renewal are the special financial characteristics that pension funds possess (Greenwich Research Associates, 1983). Pension funds are a unique source of patient growth-oriented capital. They have a long investment time horizon, which allows them to make investments that offer good returns but that are relatively illiquid over the short term. Their cash inflow from contributions and earnings will on average exceed benefit payments for sixty more years. In addition, their liabilities are only partially indexed, so they are more able to make long-term fixed-rate investments. And finally, because so many pension funds have such sizable portfolios, they are able to commit a significant dollar amount of portfolio assets to higher risk, less liquid investments, while still maintaining an overall portfolio with sufficient liquidity and a prudent degree of risk. Pension funds of states and cities average $512 million and of large corporation $250 million.

AVOIDING CAPITAL DISPLACEMENT AND FINANCIAL CONCESSIONS

Many efforts at development-oriented pension investments have unfortunately been shaped by more simplistic considerations. The most popular approach, particularly for state pension funds, involves targeting broad classes of investments simply because these investments are within the state. If the pension fund requires that these investments offer competitive rates of return, and there is no further attempt at targeting, such a practice will almost always result in investments that would have been made anyway by some other financial institution.

The reason for this is fairly simple. Despite some notable capital gaps, large segments of our capital markets do an extremely effective job of matching the users and suppliers of capital. A development investment program that targets only geographically or to broad industrial sectors is almost certain to reach only capital users that would have been well-financed by other financial institutions. Although a pension fund undertaking such a program will end up with a portfolio with a greater share of its investments in-state, providing the *appearance* of contributing to in-state development, in reality the fund will only have displaced other investors who would have financed these very same projects.

Another approach requires making financial concessions. These concessions result from an attempt to finance ventures that have favorable economic impacts but that are unable to offer competitive returns to the investor. Projects of this sort can succeed only if the pension fund accepts returns involving a financial concession. An example of concessionary investing: lending to businesses in depressed communities at below market rates of interest in the hope of stimulating employment growth.

Below market rate investments usually produce one of two undesirable results. Either contributions to the pension fund must go up or benefits must go down. Although the concessionary approach may succeed in aiding development, it surely fails to provide the best retirement benefits at the least cost. By a rough rule of thumb, a 1 percent decrease in a plan's long-term earnings rate will increase contributions—or decrease benefits—approximately 20 percent.

INVESTING IN CAPITAL GAPS

Pension funds can exert a major influence on investment patterns without making financial concessions. As mentioned earlier, inefficiencies in capital allocation can be found in many parts of our economy, and corresponding to every capital gap is a financially rewarding development investment opportunity. A few examples will illustrate this connection.

A study by the First National Bank of Boston found that more than 70,000 financially sound firms (debt to equity ratios of one or less) have found it impossible to raise medium- to long-term capital in conventional credit markets (U.S. Department of Commerce, 1981). Moderate size businesses, which are major job generators in most communities, are the primary victims of this problem. The banks and insurance companies on which moderate size businesses traditionally rely for credit have become increasingly unable to meet their credit needs. The reason is quite simple: The assets of these institutions are increasingly short-term; this makes it extremely difficult for them to make any but short-term loan commitments themselves. In addition, some parts of the country, particularly rural areas, have uncompetitive banking markets; there is substantial evidence that a lack of effective competition among leaders leads to a restriction in loans to smaller businesses, the source of most new job growth in most communities. Various state and federal banking and insurance regulations also work to the detriment of small businesses seeking capital.

Pension funds could reinvigorate these bank-starved borrowers by channeling funds through private placement debt financings or other appropriate financial intermediaries. Since 1977, the Oregon Public Employees Retirement System has purchased long-term fixed-rate commercial mortgages from financial institutions in the state. These are mainly loans to small businesses. The Baltimore municipal pension funds have recently committed $10 million to long-term fixed-rate commercial mortgage loans in the city (Saunders and Vitarello, 1983). Pennsylvania is in the process of establishing a separate account under the management of a major life insurance company; this account, to be funded by a variety of pension investors, will make long-term plant and equipment loans to medium-sized manufacturers (Litvak and Barker, 1983).

Equity financing for early stage and emerging growth companies is in equally short supply. The shortage of risk capital relative to new business opportunities has enabled venture capital investors to earn an average of 25 percent annually on their portfolios (Reinganum, 1983; Banz, 1981). Nor does the situation change once these companies reach the stock market.

A variety of recent studies show that the stock of the smallest publicly traded companies has long offered rates of return far in excess of that needed to compensate for its higher degeree of risk

(Venture Economics, 1981). Such evidence provides a good indication that obstacles exist to efficient capital flows. Although the very companies affected by these conditions are responsible for the largest portion of America's technological innovations, they are finding it exceedingly difficult to gain access to the capital they need and deserve.

This situation could be substantially rectified by the increased pension fund financing of venture capital pools and a greater demand for the stock issues of young growing corporations. For example, the Minneapolis Employees Retirement Fund, along with the Control Data Corp. pension fund, has recently invested in the Minnesota Seed Capital Fund, a vehicle providing start-up financing for the state's high technology companies. The Kansas Public Employees Retirement Fund has $80 million in a special portfolio of public equity issues of firms important to the state economy, including over-the-counter stocks. Public retirement systems in Ohio have invested $15 million in a venture capital limited partnership—the Cardinal Fund—which will make the vast majority of its investments in Ohio companies. The Michigan Employees Retirement System is a major partner in the Michigan Investment Fund, whose target is emerging technology companies in that state, along with more mature ones capable of revitalization and growth with the addition of new technology.

VEHICLES FOR DEVELOPMENT INVESTING

Pension funds have traditionally been passive, conservative investors. Many now lack the expertise and management resources for active development investing. In the case of public funds, much of this bias stems from effects of legal list constraints, which stipulate permissible investments in elaborate—and often archaic— detail. In the case of private funds, the bias has come from excessively narrow ERISA rulings that encouraged ony blue chip investments. On both fronts, however, the climate has changed markedly in recent years. Many states have chosen to liberalize the standard defining appropriate investment, and the number moving toward liberalization seems to increase each month. The climate at the Department of Labor has also thawed noticeably in recent years, and more liberal interpretations of the rules governing prudent investment standards are now relatively frequent (Litvak, 1981).

While additional reform of archaic investment standards would also contribute to development investing, legal reform alone is insufficient. Appropriate investment vehicles must be created to satisfy fiduciary concerns. And pension fund beneficiaries, trustees, and contributors must take a more active interest in how their plan's decisions affect them.

Fund managers and trustees commonly identify four characteristics as important in selecting an investment vehicle to make business development investments (apart from the obvious considerations—competitive rates of return for the relevant levels of risk).

These characteristics include (1) access to professional investment management; (2) opportunity for coinvestment with private, nonpension fund investors; (3) ability to diversify business development investments; (4) insulation from political pressure to make specific investments; and (5) opportunity to commit only a small percentage of total plan assets.

There are four different kinds of investment vehicles that can provide a way of bridging these concerns, thereby allowing funds to make development investments (Litvak and Barker, 1983).

The commingled (or pooled) fund. This is the closest approximation of the popular perception of an investment intermediary. A commingled fund pools a part of the assets of different pension funds into a single specialized investment fund with qualified independent management. The best known example of this approach is the mutual fund. The Minnesota Seed Fund, discussed earlier, is a good example of the pooled approach to development investing.

The investor consortium. In a consortium, different funds do not actually pool their money; instead, they establish a structure and a procedure for investing jointly in individual investments. An investor consortium can also provide a means of sharing certain investment advisory costs. Perhaps the best example of the consortium approach is the Development Foundation of Southern California, formed two years ago by several jointly trusteed pension funds. The foundation serves as a forum through which real estate developers, through mortgage makers under contract to the group, present investment opportunities in which the pension funds can elect to take participations.

The investment insurance program. Under this arrangement, a public or private entity pledges to cover certain investment losses in exchange for a payment (the "risk premium"). While there are other ways of transferring risk that have the effect of investment insurance, they require the existence of a substantial "capital cushion" in a commingled fund. Investment insurance is usually most helpful when applied in blanket form to pools of housing or business mortgages. Public pension funds have used private mortgage insurance to purchase more than $1 billion in targeted residential mortgages in recent years.

The specialized investment broker. Under this final option, a public or private entity acts as an intermediary between businesses seeking development capital and individual pension funds. The broker has the job of identifying appropriate deals and packaging them for presentation to pension funds. During the Brown administration, the California legislature established a state Pension Investment Unit, whose job it was to play just such a role in brokering development investments.

SOME PRACTICAL CONSIDERATIONS
FOR THE CAPITAL SEEKER

First, understand the basic nature of your development financing need. If it is simply a below market interest rate the project is looking for, look somewhere else. Pension funds are investors, not grantmakers. Consider pension funds if you are looking for an investor that would be interested in an equity participation instead of a straight loan, or in a longer term financing or fixed-rate one. (This does not mean pension fund financing cannot be combined with subsidies from other sources.)

Second, be prepared to educate and assist the potential pension fund investor. Despite their financial suitability for development investments, most pension funds only have experience investing in publicly traded securities. Identify and involve financial institutions and advisors that can assist the pension fund in evaluating, negotiating, and monitoring these investments. Such institutions will include venture capitalists, insurance companies, investment and mortgage bankers, specialized investment advisors, and banks. Any ongoing program of development investments by pension funds likely will require the establishing of one of the vehicles described above.

Third, if it is a public pension fund, be sure you understand the legal statutes governing the investments. In some jurisdictions, laws need to be updated—consider this effort a long-term investment in making pension fund capital more available.

Finally, when approaching a pension fund, make sure you deal with the appropriate person. Pension funds have administrators, trustees, pension committees, investment managers, actuaries, and a host of other personnel. The trustees have investment policymaking responsibility and the investment managers manage the portfolio on a day-to-day basis. There will usually be multiple trustees and multiple investment managers, with some of the latter in-house and some outside. If you are unable to determine who the most receptive and appropriate person would be, direct your inquiries to the chairperson of the trustees.

REFERENCES

BANZ, R. (1981) "The relationship between return and the market value of common stocks." Journal of Financial Economics 9.

Greenwich Research Associates (1983) Large Corporate Pension Plans, 1983/Public Pension Funds, 1983. Greenwich, CT: Author.

LITVAK, L. (1981) Pension Funds and Economic Renewal. Washington, DC: Council of State Planning Agencies.

———— and M. BARKER (1983) Pension Funds and Economic Development in Pennsylvania. Harrisburg: Pennsylvania MILRITE Council.

Municipal Finance Officers Association (1983) A Survey of Investment Policies and Management Practices of Public Employee Retirement Systems. Washington, DC: MFOA Government Finance Research Center.

REINGANUM, M. (1983) "Portfolio strategies based on market capitalization." Journal of Portfolio Management (Winter).

SAUNDERS, G. and J. VITARELLO (1983) "Sound investments in small business through partnerships with local banks." Government Finance (September).

U.S. Department of Commerce (1981) An Empirical Analysis of Unmet Demand in Domestic Capital Markets in Five U.S. Regions. Washington DC: Government Printing Office.

Venture Economics (1981) Venture Capital Investment Analysis. Prepared for the Office of Economic Policy, Planning and Research, Department of Economic and Business Development, State of California.

Part IV

Local Economic Development at Work

ox why ecoder works in Beltimar
why wasn's so for in Detroit

Economic Development Really Works: Baltimore, Maryland

BERNARD L. BERKOWITZ

☐ THE INCREASINGLY FAMILIAR picture of Baltimore's new and exciting Inner Harbor waterfront with its Harborplace pavilions, National Aquarium, and World Trade Center and with its gleaming new office and hotel buildings in the adjacent central business district has come to symbolize the renaissance of Baltimore and the success of its economic development program. While the Inner Harbor project is Baltimore's most dramatic economic development achievement, and one of tremendous substantive as well as symbolic importance, it is only one component of a comprehensive, and by and large successful, economic development program. Baltimore has certainly not solved all of its economic problems—no older city has—but economic development really works in Baltimore.

This chapter describes the accomplishments in Baltimore during the past two and one-half decades. While the focus is on projects and programs, these are presented within the context of the city's geographic, economic, social, and institutional situations, and it's problems. Particular emphasis is given to the institutional and organizational factors responsible for Baltimore's sucess: (a) a comprehensive and multifaceted economic development strategy; (b) the city government's partnership with the private sector; (c) the creative use of quasipublic development corporations and other innovative approaches; and (d) the strong mayor form of government under a dynamic and dedicated Mayor, William Donald Schaefer.

BACKGROUND

Baltimore is the furthest west of the major east coast ports and is located between the North and the South. The favorable fact of geography led to Baltimore's growth as a port industrial and commercial center and is a continuing source of economic strength.

The city's population was about 780,000 in 1980 as compared to a total of 2,250,000 in the metropolitan area (the city and five surrounding counties).[1] Population in the city has declined by 18 percent since 1960 while the area has increased by 50 percent.

Total employment in the city is just under one-half million, nearly half of the metropolitan area total. The May 1983 unemployment rate was 8.9 percent for the city and 7.6 percent for the area, both figures below the national rate.

Baltimore's resident labor force is large and diverse and contains many persons with specialized skills that are still important to many local industries, such as specialized machinery manufacturing. In general, the labor force has a good reputation for skill, productivity, and cooperation.

By the end of the 1950s, Baltimore was experiencing the same kinds of problems as most older American cities—obsolete declining commercial and industrial districts; loss of businesses and jobs mainly to the suburbs; shortage of vacant land for industrial sites; flight of white midde-class residents to the suburbs; a stagnant tax base; and so forth. In the 1960s and 1970s, it also became more apparent that the structural shift in the national and area economies from manufacturing to service industries and from blue-color to white-collar activities was producing a serious mismatch between the city's resident labor force, particularly persons with lower educational attainment, and the kinds of jobs available in the growth sectors. The current "displaced worker" problem is one serious manifestation of the changing composition of the economy, another being the difficulty of absorbing unskilled and undereducated young entrants in the labor force.

By 1970, disparities between the city and its largely affluent metropolitan counties also became more apparent—that is, higher unemployment, a disproportionately large population reliant on public assistance, and relatively low per capita income and tax yield.

BALTIMORE RESPONSE

From the late 1950s to the present, there were many events and initiatives by the public and private sectors that had a positive and reinforcing impact on Baltimore's economic situation. However, two stand out.

The first was the private sector initiative in 1958, through the Greater Baltimore Committee and the Committee for Downtown, proposing the first major downtown redevelopment project, Charles Center. The acceptance and implementation of that plan by the city brought into being the continuing public-private partnership so important to Baltimore's renaissance. It also led to other downtown redevelopment, particularly the Inner Harbor project, and had a salutary effect on public and developer psychology regarding city redevelopment.

The second was the election of William Donald Schaefer as Mayor in 1971, followed in the early 1970s by the formulation and implementation of a comprehensive development strategy that included new initiatives in industrial development; the revitalization of older neighborhoods, including their retail areas; and attracting more conventions and tourism, as well as continuation of the successful downtown redevelopment program.

The first initiative got Baltimore's renaissance started and its success made other development initiatives credible. The second triggered a wave of closely integrated initiatives that moved Baltimore from a limited project orientation to a comprehensive economic development strategy. Because of their significance, the two are described in detail below.

THE GREATER BALTIMORE COMMITTEE AND THE CHARLES CENTER PROJECT

Concern about the obsolescence and decline of the central business district (CBD) and the city's fiscal situation led a group of progressive business leaders to form the Greater Baltimore Committee (GBC) in 1956.[2] (Under its able and aggressive executive director for all but the last two years, William Boucher III, the GBC has played a large and positive role, not only in the downtown redevelopment program and economic development, but in a broad spectrum of civic issues in concert with the city government.) The Planning Council of the GBC took the unprecedented step of hiring its own professional staff. Although the initial emphasis was on an overall plan for the CBD, there was early recognition that positive action was needed quickly to stem the reality and psychology of decline. This led to the preparation of a dynamic plan to redevelop 33 acres of land at the heart of the CBD, Charles Center.

Charles Center. This plan was first presented by the GBC and the Committee for Downtown to the then Mayor, Thomas D'Alesandro, Jr., in March 1958. Later that year, the city agreed to undertake Charles Center as an urban redevelopment project. That was quickly followed by the authorization of a city general obligation bond issue by the Maryland General Assembly and the city's voters in 1959 and by the approval of the urban renewal plan by the Mayor and City Council in March 1959.

The project has been implemented by the Charles Center Inner Harbor Management, Inc., the first of Baltimore's highly effective nonprofit, quasipublic, development corporations. Originally (1959-1965) called the Charles Center Management Office, the corporation operates under a contract with the city through the Department of Housing and Community Development and takes its policy direction from the city administration. The general manager of the predecessor Charles Center Management Office and first chairman of the Corporation was the late J. Jefferson Miller, a retired department store executive and a $1-a-year man with Charles Center. The president of the corporation is Martin Millspaugh, who joined the organization in 1960, a year after Miller. Walter Sondheim, Jr., also a retired department store executive and civic leader, succeeded Miller as chairman in 1972. The quasipublic development corporation and its private sector setting and style was a key to encouraging private investment and strengthening the public-private partnership.

Public expenditures of $38.8 million, including $10.5 of city bond funds and $28.3 of federal Title 1 redevelopment and other funds, provided the setting in the form of plazas, fountains, overhead walkways, parks, landscaping, and so on, and the cleared sites. This has attracted $145 million of mostly private development including over 2 million square feet of net rentable office space, 700 hotel rooms, 700 apartment units, 430,000 square feet of retail space, 4,000 parking spaces, and an 1,800-seat legitimate theater. Five existing structures were retained and improved. Only two sites are left in Charles Center. An office building is under construction on one and an apartment-commercial development is committed on the second.

Employment in Charles Center is estimated at about 15,000 as compared to 9,800 before redevelopment. Real property tax revenues in fiscal year 1981-1982 were $2.4 million, over three times as much as before redevelopment. Furthermore, the high quality of design and public amenities in Charles Center are an attraction to visitors and tourists.

Clearly, Charles Center is an economic development success. It was also important for the psychology of Baltimore's citizens, including developers, because it dispelled the skepticism that prevailed at the start of Charles Center. Success in downtown redevelopment led to the anticipation of further successes and provided momentum for the larger, and now more important, Inner Harbor project, which is discussed later.

MAYOR SCHAEFER AND THE CITY'S COMPREHENSIVE ECONOMIC DEVELOPMENT STRATEGY AND PROGRAM

Baltimore's strong mayorial government gives the mayor initiatory powers over the budget and the appointment of department heads. This provides greater potential for the coordination and consistency of city programs and projects. William Donald Schaefer has turned potential into reality through the creation of a cabinet structure, personal involvement in city programs, the encouraging of innovation, and the demanding of results.

Before becoming mayor, in December 1971, Mayor Schaefer's experience as President and Member of the City Council had resulted in considerable knowledge and interest in economic development, small business, and neighborhood retail areas. His early involvement with the Citizens Planning and Housing Association was also reflected in a commitment to strong residential neighborhoods and housing programs. This deep and varied background was conducive to providing leadership for a comprehensive assault on the city's problems.

The economic development component of Baltimore's comprehensive plan was published in 1973 by the City Planning Commission, of which the Mayor is an ex-officio member. (Mayor Schaefer's representatives on the commission have been his physical development coordinator, Mark Joseph, Mark Wasserman, and myself.) This document set forth a series of interrelated goals, objectives, and policies for the city's physical and human resources development as related to economic development. It ranged from broad policies as, for example, capitalizing on the location and design of the interstate highway and transit systems to maximize economic development potential, to specific proposals for industrial parks and financing programs.

The City Charter of Baltimore was amended in 1964 to provide for a six-year capital improvement program to be adopted by the Board of Estimates, the city's five-member governing body, upon recommendation of the Planning Commission and the Director of Finance. Significantly, the first officially adopted capital improvement program occurred in 1972, the first year of Mayor Schaefer's first term, and the Board of Estimates has adopted a six-year program (CIP) every year since in connection with the annual capital budget. The CIP process is used creatively as a means of coordinating and managing the comprehensive development program—that is, as a tool for implementing the comprehensive plan. The economic development component of the comprehensive plan has been the framework for the city's many successful projects and programs.

THE STRATEGY AND ITS COMPONENTS

The economic development strategy is itself part of a larger comprehensive development strategy that includes the revitalization of the city's residential neighborhoods through both new and rehabilitated housing, improved community facilities and services, and a partnership between the city and its communities; a balanced set of new and improved transportation facilities including the interstate highway system and a subway, to improve access to downtown and the major port and industrial areas and to reduce through traffic in neighborhoods; and greatly strengthened programs, particularly education and manpower training, to enhance human resources development.

The components of the economic development strategy are discussed below in two groups: those for which the Baltimore Economic Development Corporation is responsible and those for which other organizations are responsible.

THE BALTIMORE ECONOMIC DEVELOPMENT CORPORATIONS (BEDCO)— INDUSTRIAL RETENTION AND ATTRACTION

Mayor Schaefer's platform in 1971 included the creation of a new economic development program to establish an industrial land bank and develop new industrial parks for the purpose of strengthening industrial retention and attraction to Baltimore. More specifically, he supported a $3 million bond issue that had been defeated by the voters in 1968 and was put back on the ballot in 1971. The mayor gained the support of the GBC, the Sun newspapers, and other organizations for the new program. With that support and effective communication of the benefits to community groups, the $3 million bond issue was approved by the voters in 1971. Subsequent bond issues to continue capital funding of the program have been approved over the years.

The approval of the bond issue led to the creation of BEDCO,[3] a one-step service to businesses seeking a plant location (site), financing, technical services, and assistance in dealing with building and other governmental regulations and services. Since 1976, BEDCO's development projects and financing assistance have been instrumental in the retention or creation of approximately 24,000 jobs in Baltimore and additional real estate taxes of over $3 million per year. This does not include the firms for which BEDCO has provided ombudsman services or helped to locate on privately owned sites or in privately owned buildings.

BEDCO's land banking and industrial park development program. This program has resulted in the creation of six city-owned industrial parks in Baltimore with an aggregate of 360 acres. Vacant and underutilized land bypassed by private development has been acquired; derelict buildings cleared; streets and utilities installed; and land marketed by BEDCO. An eight-story 400,000 square foot vacant former clothing manufacturing plant has also been renovated and a new garage built to create a new "vertical industrial park," the Raleigh Industrial Center. In addition, the city's commitment to environmental improvements through urban renewal action was instrumental in the successful development of a 40-acre privately owned business park, Caton-95 in southwest Baltimore.

The new industrial parks have enabled the city to be "in the game" by having an ongoing inventory of available sites. Of the approximately 400 acres in the seven industrial and businesses parks, about 230 have been sold, leased, or earmarked and the Raleigh Industrial Center is about three-fourths leased. These projects now contain 60 firms with an aggregate employment of 3,100 and private investment of $62 million. The added real property tax receipts are about $1 million per year. While the 60 companies are mostly city firms that had outgrown previous sites, there are a number of firms that moved into the city from the counties or are completely new establishments—for example, Hewlett-Packard at Caton-95 and Hitachi Metals at the Bayview Industrial Park.

The city's industrial parks range from the fully developed 19-acre Crossroads Industrial Park (former stockyards) to the 170-acre Holabird Industrial Park (formerly the Army's Fort Holabird acquired from the federal government). The others are Quad Avenue (totally committed) with 24 acres, Bayview, Park Circle, and the Seton Business Park. Upon full completion, the six city industrial parks, the Raleigh Industrial Center, and Caton-95 are expected to have firms with 8,000 to 8,500 jobs, private investment of $110 to $120 million, and real estate taxes in excess of $2 million per year.

In developing the industrial parks and the Raleigh Industrial Center, Baltimore has greatly benefited from more than $18 million in grants for infrastructure and building improvements (Raleigh) from the Economic Development Administration (EDA) of the U.S. Department of Commerce. In addition, $7 million in City General Obligation Bond Funds and nearly $5 million from a portion of the proceeds of the sale of Friendships Airport to the state has been used for property acquisition and improvements. Another important source of funds has been loans (up to $1.5 million per project) under the State of Maryland's Industrial Land Act (MILA) program.

In most cases, the city has been able to recover its investment and pay off the state loan in BEDCO developed projects from the sales and lease proceeds. The Crossroads Industrial Park, for example, resulted in $1.1 million of land sale proceeds, enough to pay off the MILA loan and recover the city's investment. In addition, the project resulted in 400 jobs and an increase in real property taxes of $160,000 per year.

Within the next two years, about 150 acres of land in new or expanded city industrial and business parks will be added to the program, including a high priority biomedical business park.

Improvements to older established industrial districts are important for industry retention, as are infrastructure improvements for individual firms. The most comprehensive improvements have been made in the Carroll Industrial Park, an existing industrial area just southwest of the CBD and on an entranceway from the airport. There are about 50 companies and 5,000 jobs in the area. About $2.25 million of city funds and a like amount of EDA funds have been invested in the reconstruction of streets and utilities, planting of trees, and creation of off-street parking. BEDCO also encouraged the companies to organize a steering committee; assisted with the design of facade improvements, expansion, and problem solving; and helped finance investments in the area. The name Carroll Industrial Park has caught on, the area has been stabilized, and significant private investment has occurred.

The retention and expansion of existing businesses. The retention and expansion of businesses (small, medium and large) has been a high priority of BEDCO. It is BEDCO's most important outreach and marketing activity, which takes a number of forms.

Mayor Schaefer makes about six visits to major plants each year. These visits are carefully organized by BEDCO in advance of the visit. A detailed agenda is prepared with the plant manager covering all issues important to the company and/or the city. The appropriate city agency heads accompany the mayor. The visit consists of a meeting to discuss the issues—such as street repair, clarification of city plans, the company's need for financing to expand, security, and so on—and a tour of the plant. The Mayor assigns follow-up responsibilities to city staff and BEDCO is responsible for coordinating follow-up.

Other major plants are visited by top BEDCO staff each year. In addition, hundreds of letters are sent to smaller companies telling them that we are glad that they are located in the city, advising them of BEDCO's services, and offering assistance with problems or expansion. These letters are followed up by phone calls and staff visits to the firms.

The retention program also utilized advice and assistance (including early warnings of problems) from the Growth Assistance Committee of the Economic Development Council of the GBC and from an interagency committee of city officials.

Major successes of the retention program include the decision by General Motors to modernize their city assembly plant at a cost of $270 million, and the current investment programs of Armco Steel, Lever Brothers, Amstar (Domino Sugar), and so on. Infrastructure improvements, making land available, financing assistance (including financing acquisitions by local management), and general ombudsmanship have all played a part. The result has been a more favorable record in retaining industry than has been true in many areas of the Northeast and Midwest.

The growth of new small businesses. This form of growth has been encouraged through special development projects (incubator buildings). The formation and growth of new businesses has been documented by studies that have shown that small business accounts for most of net employment growth and innovation in the economy.

The Raleigh Industrial Center has functioned as an incubator of small enterprises as well as a relocation resource for small firms being displaced by CBD expansion. There are over 20 companies and 400 jobs in the building; all are small; many are completely new to the city or were newly formed; and one-quarter are minority owned. The excellence of the space and facilities (loading, elevators, etc.) combined with economical rents, parking garage, convenient location, and city financing of leasehold improvements have been attractive to the tenants.

The Raleigh Industrial Center project involved about $8 million in capital expenditures, nearly $5 million from EDA. This includes acquisition, modernization of the building, construction of a 250 car garage, and leasehold improvements. The center opened in 1980 and is now three-fourths leased.

Another incubator building in Baltimore is the Control Data Corporation's Business and Technology Center (BTC) recently completed in the city's Park Circle Industrial Park. The city leased a vacant former school to Control Data Corporation, which renovated the school as office space for tenants and for Control Data's services to tenants. In addition, the nearly 6-acre site was used to build a 50,000 square foot addition on the school for industrial tenants.

The BTC has achieved an occupancy level of 54 percent in less than 6 months. There are already 41 tenants and 200 jobs in the building with most employees coming from the Park Heights neighborhood. Most of the companies in the building are small and minority owned. The BTC offers tenants a comprehensive array of services including answering services, duplicating, secretarial, 24-hour security, access to a computer, financing, and so on.

Later in 1983, construction will start on renovating a vacant 30,000 square foot buiding at the City Hospital into a incubator for biomedical firms. A new biomedical firm with close ties to the Johns Hopkins Hospital in the research phase will occupy one-third of the space and employ about 30 persons.

It is also important to note the contribution of financing assistance, including the leveraging of city loans and the use of the SBA 502 and 503 programs, to the growth of small firms.

Financing incentives. These make up an essential part of the city's economic development program. Baltimore's voters have on numerous occasions authorized the use of general obligation bond issues for industrial development financing, commercial development financing, housing financing, and developer financing. In addition, the city has made great use of the Federal Urban Development Action Grant (UDAG) Program for attractive financing, industrial revenue bonds, the SBA and EDA loan programs, and a number of State of Maryland programs. Creative financing has often made a critical difference for projects, particularly because Baltimore does not have the power to abate real property taxes (except for state enterprise zones) and does not offer grants to firms.

In order to administer the city's loan programs and advise the city's Board of Estimates on loans, that board delegated certain responsibilities to an entity known as the "Trustees for Loans and Guarantees." The two trustees are Charles Benton, the city's Director of Finance, and Frank Baker, the former head of the Monumental Life Company and a respected businessman and civic leader. The trustees have their own staff supported from loan fees and revenues. Loan or loan guarantee proposals are recommended to the trustees by BEDCO, CCIH, MCDC or HCD. The trustees analyze the proposals, sometimes modify them, and recommend approval or disapproved by the Board of Estimates, the decision-making body. Once approved, the trustees administer the loan or guarantee.

Since 1976, BEDCO has initiatied $540 million in financing, mostly through bonds, also a variety of federal, state, city, and private sources, including the BEDCO Development Corporation as a direct lender under the SBA 503 program.

More often than not, BEDCO packages two or more loan programs to leverage city funds and provide a competitive composite. Some examples of this are (a) a new foundry located in the city's Bayview Industrial Park instead of southern Pennsylvania by using a state MILA loan to finance the building and revenue bonds for the machinery and (b) a textile manufacturing firm moved with its 280 jobs from a flood-prone county site to a 300,000 square foot building in the city instead of going to South Carolina as a result of a $6 million package including a $2.5 million IRB, an $850,000 UDAG, a $2 million state loan, a $200,000 trustees loan, and $.5 million equity.

Baltimore's ability to capitalize on special economic development opportunities has been an important element of its strategy. Two examples are Baltimore's foreign trade zone and the state designated enterprise zone.

As one of the nation's largest ports and a center of manufacturing and distribution, a foreign trade zone (FTZ) is a natural for Baltimore. However, Baltimore was interested in the job intensity that would result from assembly and manufacturing, not in the warehousing that characterizes so many U.S. FTZs. Thus, it was only after the law was changed by Congress to exclude value added by labor from duties within the zone that BEDCO applied for an RTZ in 1981 for 20 acres at the Holabird Industrial Park.

In January 1982, Mayor Schaefer announced the federal approval of the FTZ and the selection of the developer and operator of the zone. An attractive 58,000 square foot multitenant building was completed in 1983. Upon full buildout, the FTZ will contain 350,000 to 450,000 square feet of space, provide 500 to 800 jobs, and produce tax revenues of about $350,000 per year. The city will also share in net profits.

Although Congress has yet to enact the proposed federal enterprise program providing targeted tax incentives and regulatory reductions in zones, Mayor Schaefer and BEDCO initiated state legislation that was supported by Governor Harry Hughes and by all parts of the state and was enacted by the Maryland General Assembly in 1982.

The Maryland program provided for up to six zones to be designated each year, but no more than one for a political subdivision. Firms located within an enterprise zone are entitled to an 80 percent reduction in real property taxes for five years on new investment or additions, sizable credits against their state income tax liability for net additional persons hired, and access to a special enterprise zone loan fund. The political subdivision is to be reimbursed by the state for 50 percent of the real property tax abatement and also has greater access to state loans and grants for projects within the zones.

In December 1982, four state zones were designated, including the city's Park Circle Industrial Park. Baltimore has also applied for two other zone designations.

Mini-industrial parks in low-income neighborhoods provide a means of targeting job creation to those most in need. The Park Circle Industrial Park, for example, located in northwest Baltimore in the Park Heights area, is a large predominately black community with about 45,000 persons. Park Heights is also the city's largest urban renewal project and has benefited from millions of dollars spent on housing rehabilitation of its basically sound stock, new schools, recreation centers, multipurpose centers, parks, playgrounds, educational facilities, and other services. Despite these improvements, the unemployment rate is over twice the city average and crime, delinquency, and other related social problems continue. The result has been a strong community interest in economic development, including the formation several years ago of a capable community development corporation, the Park Heights Development Corporation.

The existence of vacant and underutilized land at Park Circle presented the opportunity to develop jobs in the community. This was recognized in the city's plans and received strong support from the community, Congressman Parren J. Mitchell, and the mayor.

In June 1980, Mayor Schaefer announced the start of the Park Circle Industrial Park and the employment of the City Venture Company as consultant on overall project strategy and marketing. The mayor also appointed an advisory committee with neighborhood, business, and government representatives. The goal for the project was set at 1750 jobs with another 750 to be located elsewhere in Park Heights.

The Park Circle Industrial Park is another example of the Baltimore partnership principle, in this case between the community (particularly the Park Heights Development Corporation and the Park Heights Community Corporation), the private sector (particularly Control Data, Commercial Credit, City Venture, and the Private Industry Council [PIC]), and the city (BEDCO, HCD, and the mayor's Office of Manpower Resources.)

About $6 million from various federal, state, and city sources has been allocated to the project for property acquisition, clearance, and infrastructure improvements. Property acquisition is complete; demolition nearly so. Streets and utilities will be finished at the end of 1983, and landscaping and sodding are partly done. As already noted, Control Data has built a Business and Technology Center that is 54 percent leased and already has 200 people working there, including 60 in a bindery established by Commercial Credit. In addition, a new commercial food equipment manfufacturing plant is now under construction and will initially employ 60 persons. A "job match" system has been established for Park Circle. This involves a commitment by occupant firms, expressed in disposition agreements, to utilize the mayor's Office of Manpower services for training and placements with focus on neighborhood people. A sign at Park Circle says "something special is happening here." The project is off to a good start and was visited by President Reagan in July 1982.

Park Circle is the prototype. Other similar projects are underway or planned in low-income distressed neighborhoods.

Assisting the formation and growth of minority owned businesses is a priority for Baltimore. It is part of a commitment to equal opportunity, affirmative action, and the goal of equity. It is also based on the greater likelihood of black-owned firms locating in black neighborhoods and hiring disadvantaged blacks.

Minority entrepreneurs have been encouraged to acquire viable majority-owned businesses. For example, after a careful analysis by BEDCO, the city acquired a vacant concrete manufacturing plant and offered it to minority entrepreneurs. The plant was acquired by an existing minority-owned concrete fabrication company with city financing. A manpower program contract also provided on the job training subsidies. The plant is now a successful profitable operation.

Financing assistance has been important in the formation and growth of minority-owned businesses in Baltimore. Recently, the city's commitment to minority enterprise was further illustrated by granting the new Development Credit Fund (DCF) half of its first year start-up operating expenses. The DCF is a state-guaranteed loan fund with money ($15 million) made available by Baltimore banks for working capital and machinery—equipment loans of up to $500,000 for minority-owned firms. With bank and minority community representation on the DCF Board and Loan Committee, this is a significant advance in resources and commitment to minority enterprise, and another example of public-private partnership.

Marketing and promoting Baltimore to the rest of the nation and world. This is important for Baltimore to attract new investment. However, the city, the private sector, and BEDCO have determined that it makes more sense to have a coordinated effort to market the entire metropolitan area rather than for the city and each of the counties having separate uncoordinated marketing programs. Thus, the city and the counties contribute modest funding for a largely private sector funded promotion and marketing program of the Economic Development Council (EDC) of the GBC. This is done in close cooperation with the marketing program of the Maryland Department of Economic and Community Development (DECD).

Four years ago, BEDCO used city funds to create the Baltimore Briefing Center, a place with central information, audiovisuals, models, and so on, demonstrate the advantages of Baltimore to prospects. Three years ago, BEDCO contracted with EDC to operate the center. It is now funded under a partnership agreement of the city (BEDCO), EDC, and the state (DECD), with policy direction from the heads of the three organizations.

BEDCO's marketing efforts, as previously noted, are focused on existing Baltimore businesses. The more limited independent external marketing is carefully targeted on selected opportunity markets. In particular, a concerted effort is being made to attract biomedical firms. Considerable help is being provided by key staff of The Johns Hopkins Medical Complex. Hopkins and the city have a common interest in developing a significant medically related industry concentration in Baltimore. Hopkins and the University of Maryland are important magnets for this kind of business from both a research and procurement point of view. The potential for enlarging the city's industrial base through capitalizing on this high technology field has already been noted.

NON-BEDCO PROGRAMS

Continuation of the successful Downtown Redevelopment Program. This continuation retains a high priority. Charles Center and the Inner Harbor have enabled Baltimore City to capture one-half to two-thirds of the metropolitan office space demand during the past two decades. As a result, Baltimore has benefited from growing office, business service, financial, and the like employment and real property tax base. It has also been essential to the development of jobs and tax base from conventions and tourism.

(a) *The Inner Harbor Project* continued the partnership between the City and the private sector as GBC and the Charles Center Inner Harbor Management Corporation cooperated in the funding of the plan that was prepared by David Wallace's consulting firm. Unveiled in 1964, it covers 240 acres, divided subsequently into several separate official renewal plans. Implementation began in the late 1960s.

As in the case of Charles Center, public investment in acquisition, clearance, new bulkheading, a continuous promenade, docks, marinas, open space, overhead walkways, and so on, provided the sites, amenities, and setting for private development. Public funding for these purposes has amounted to about $100 million including $75 million in federal grants, $17 million in city bond funds, block grant funds, state grants, and so forth. Buildings representing about $1.5 billion in new private, public, and institutional investments are either complete, under construction, or committed. This includes the nationally acclaimed Rouse Company Harbor Place projects, the National Aquarium, World Trade Center, and many others. Another $.5 billion is planned. The 1981-1982 real property tax yield was $4.1 million with the ultimate increase expected to reach $15 to 20 million.

(b) *The Market Center Area Project is the most recent major downtown redevelopment project.* This area is the old main downtown retail district west of Charles Center, east of the University of Maryland Baltimore City Campus, north of Inner Harbor West, and south of the Seton Hill and Mount Vernon residential areas. The focal point of the area is the old "one hundred percent" corner of Lexington and Howard Streets.

Like the earlier Charles Center and Inner Harbor projects, the plan involves major public improvements around which private investments will occur. These include:

— Opening in the fall of 1983 of the subway (northwest transit line) and particularly the Lexington Market Station and adjoining plaza, a generator of large pedestrian movements and an open space amenity for the Murdock Company's Atrium Project.

— Extension of the existing Lexington (pedestrian) Mall between Howard and Eutaw Streets, scheduled to be completed by December 1983.

— Construction of an attractively landscaped transit mall on Howard Street with the first two blocks of the planned 10-block mall to be finished by the fall of 1984.

— Completion in October 1982 of a $6 million addition and $2.5 million (including $1.5 million from EDA) renovation of the Lexington Market.

Implementation of the Market Center project is the responsibility of the newest of Baltimore's quasipublic development corporations, the Market Center Development Corporation (MCDC).

The mayor and city council passed an ordinance for the Market Center Urban Renewal area. While providing sites for major new office-retail development, the emphasis is on rehabilitation of more than 500 existing buildings in the area. The plan requires completion of facade rehabilitation by July 1984 and this is well under way. The mayor and city council have also enacted an ordinance establishing a special assessment district to fund an area business association that will be responsible for unified promotion, marketing, and maintenance.

The major new private investment committment in Market Center is by the Murdock Development Company (California). Murdock's first phase is the Atrium at Market Center, now under construction, and encompassing the rehabilitation of a historic building and a new seven-story 115-000 square feet office-retail building, integrated with the design of the adjacent Lexington Market subway station. Future phases include about a million square feet of new office space.

Fostering a growing convention and tourism industry. This concept has been a key successful component of Baltimore's strategy.

The Baltimore Office of Promotion and Tourism (BOPT) was created eight years ago, and has been under the leadership of its aggressive and creative director, Sondra Hillman. Its initial emphasis was on information and special events designed to attract people from the Baltimore metropolitan area to downtown. The success of this initial goal and the increasing number of tourist attractions such as Harborplace and the Aquarium led to a growing emphasis on the external promotion of tourism.

The free special events created and promoted by BOPT include ethnic festivals, "Sunny Sundays" arts and crafts exhibits, the On Stage Downtown series of concerts, Midtown music concerts, the Farmer's Market, Preakness Week Festivities, and the like. These events brought hordes of persons to the Inner Harbor, Hopkins Plaza, Market Place, and so on. In fact, the large crowds coming to events in the Inner Harbor influenced the Rouse Company to build Harborplace, which, of course, has strongly reinforced the attraction of the Inner Harbor.

Another important reason for Baltimore's emergence as a place to visit is its new $45 million convention center and the convention bureau that markets Baltimore for conventions.

The impact of Baltimore's new convention and tourism industry is indicated by an estimated 16,000 jobs and $184 million payroll in 1981. Another indication of the impact is current and planned construction of more than 2,000 additional hotel rooms.

The revitalization of existing older neighborhood retail districts is a responsibility of the Department of Housing and Community De-

velopment (HCD). This program was created in 1975 as a result of previous studies and efforts to improve some of the older strip retail districts. Studies had shown a large decrease in businesses, jobs, and assessed value within most older retail districts during the 1950s, 1960s, and early 1970s because of obsolescence, deterioration, and the competition of shopping centers.

For an area to be included in the program, the merchants must be organized and committed to investing in improvements to their properties. The affected community must also be supportive. There must be as well a reasonable prospect of success based on market potential and condition of the district. Seriously dilapidated areas require more drastic redevelopment than is characteristic of the commercial revitalization program.

When the above conditions are met, HCD staff work with the merchants and community to formulate a plan for selected public improvements and district rehab and design standards. This plan is embodied in an urban renewal plan ordinance adopted by the mayor and city council.

Private rehabilitation standards encourage facade improvements, uniform signing, building repair, and the like. Loans, generally up to $60,000 per property, are made available for the improvement of a property. These are often included in a larger loan package including the SBA 503 or 502 programs. Private owners are given two or three years to fix up their properties. The public funds leverage a multiple of private investments. Public improvements may include new streets, sidewalks, malls, street furniture, trees and landscaping, plazas, off-street parking, and so on.

Completed projects have either stabilized the improved districts or resulted in significant growth in businesses, jobs, and assessments. The most successful projects include the Oldtown Mall (the country's first mall in a predominately black, low-income, inner-city neighborhood, a project started before 1975); Mount Washington; Washington Boulevard; and York-Woodbourne, part of the York Road Corridor where the York Road Planning Area Committee (YORKY) enlisted city and private support to establish a staff with planning and financing capability. As a result of this unique partnership, millions of dollars of private and public funds have been invested in the York Road Corridor. Furthermore, YORKY has provided a model for other newer corridor development efforts.

Another HCD program has been used in conjunction with commercial revitalization. That is the shopsteading program. Modeled after Baltimore's successful "dollar house" urban homesteading program, shopsteading makes available vacant properties for only $100 to investors that will rehabilitate the properties and establish their businesses there. Dozens of such properties have been improved particularly on portions of East and West Baltimore Street.

Education (particularly vocational) and manpower training. These are also part of the overall economic development strategy. These programs determine the quality of the labor force and, therefore, the ability to attract and retain jobs in the growth sectors (generally technology and skill oriented) now and in the longer run, and the ability of the labor market to adjust to changes in composition of the employment.

Given a large segment of the existing labor force that is less educated and unskilled, it is necessary now and in the short run to retain as many blue-collar jobs as possible. However, at the same time, the education and manpower training systems must be preparing the labor force (especially new entrants) to meet the educational and skill requirements of the future.

The city administration has given special priority to education and skill development. This has been reflected in marked improvements of reading and mathematics test scores compared to national norms. The private sector has been involved through the "Adopt a School Program"; a strong emphasis has been placed on basic skills and on computer literacy; a high school of the performing arts was opened; and a new state-of-the-art Westside Skill Center was built. Special programs funded through the manpower program have been developed to provide school work opportunities for dropouts and potential dropouts.

Close Coordination of the economic and community development programs with manpower training and placement. This action is a conscious part of Baltimore's development program. Economic developers regularly make use of labor training subsidies as incentives to attract new industries and encourage existing businesses to expand. BEDCO has likewise used the state's training program and/or the mayor's Office of Manpower Resources (MOMR) programs such as on the job training subsidies to encourage growth in Baltimore. MOMR on its part has not only looked to BEDCOs, HCDs, and other development projects as a source of new job slots for job placement of MOMR's clients, but has collaborated with Baltimore's development agencies to foster economic and community development that produces a net increase in job openings targeted to MOMR clients.

Through its Private Industry Council's "Starters" program, MOMR has provided the funding for additional loan packagers in BEDCO and HCD on the basis that additional financing assistance staff would produce more loans for the growth of small- and medium-sized businesses, that would lead to an increase in job openings. Targets were established for the increase in job slots and the funding is proportionate to results. BEDCO and HCD on their part agreed to make the loans subject to the lenders' willingness to utilize MOMR's training and placement services. BEDCO has extended the requirement of the company's commitment to coordinate recruitment and training with MOMR to industrial revenue bond financing and all other BEDCO financing activities. Development projects like Harborplace, the Hyatt Hotel, Park Circle Industrial Park, and the like have provided thousands of slots for MOMR.

The Private Industry Council has also funded a small managerial and technical assistance program for small firms as well as the first two years of the City Venture Corporation contract for the Park Circle Industrial Park. As already noted, the job match program created for the Park Circle Industrial Park provided a carefully developed procedure to target jobs in that project to Park Heights residents through MOMR.

The latest example of creative BEDCO-MOMR cooperation is a government procurement outreach assistance program with funding from Title VII CETA money. This program in cooperation with Baltimore County and the EDC-GBC will provide training for economic development staff in the procurement process with the trained staff reaching out to firms to make them aware of procurement opportunities and to assist the firms to obtain more government contracts.

The cooperation between MOMR and the development agencies did not happen overnight. It evolved over a number of years and was stimulated by the interaction of the city's cabinet structure and participation in the late 1970s in the HUD, EDA, and Department of Labor demonstration grant program to foster coordination of economic development, community development, and manpower training programs.

CONCLUSION

Baltimore economic development is a success story; the reasons for which are both institutional and the result of the creative individual talents noted in this chapter. Much remains to be done, but the basis for future success has been established. Perhaps the most difficult task will be the formation of new institutional arrangements—that is, the creation of new partnerships between public, private, and university sectors and the strengthening of university resources—needed for Baltimore to capitalize on its potential growth in medically related industries and other selected technology-based industries. Based on the creativity and flexibility that has characterized Baltimore's economic development, there is good reason for optimism about this particular development thrust and the future generally.

NOTES

1. Baltimore City does not lie within any county but has the status of a county in Maryland.

2. The Chamber of Commerce of Metropolitan Baltimore was merged into the Greater Baltimore Committee on January 1, 1978.

3. The Baltimore Industrial Development Corporation, predecessor to BEDCO, existed along with the City Economic Development Department between 1972 and 1975. The functions of both were merged into BEDCO, which began operating in January 1976.

12

Economic Development or Economic Disaster?: Joliet, Illinois

CLAIRE L. FELBINGER

☐ IF ONE WOULD characterize Baltimore as an economic development success, one would have to classify Joliet, Illinois as a disaster. Joliet gained national prominence in August, 1982 when the Washington *Post's* business section contained a front-page article suggesting that the only activity thriving in Joliet was the license plate business—at the Stateville Correctional Institution. The *Post* article was in reaction to the release of the July, 1982 unemployment figures that gave Joliet the dubious honor of having the highest unemployment rate of any urbanized area in the country—25.2 percent.

The purpose of this chapter is to determine the extent to which this unemployment rate was an indicator of short-term forces beyond the control of the city and the extent to which this performance was directly linked to the city's political, social, and institutional makeup. The conclusion is that both short-term forces and long-term internal conditions were operative. This case study provides a stark contrast to the previous chapter. In all fairness to middle-sized cities,[1] however, one would not necessarily expect a one-to-one correspondence between the scope and impact of activities undertaken by Baltimore and those appropriate to a city the size of Joliet. But it is interesting to compare the causes of the disparities and their effect on the city's economy.

AUTHOR'S NOTE: *Data for this chapter were gathered from public records and extensive interviews with political and economic leaders from Joliet and private citizens. I would like to thank Anthony J. Uremovic, Laura L. Moser, Irene S. Rubin, and the editors for their insights and comments on earlier versions of this chapter.*

BACKGROUND

Joliet is located forty miles southwest of downtown Chicago at the intersection of Interstate Route 55 (Chicago to New Orleans) and Interstate Route 80 (New York to San Francisco). The DesPlaines River, part of the Illinois Waterway System, runs through the center of the city providing an all-water connection to the Great Lakes, St. Lawrence Seaway, the Mississippi River, and the Gulf of Intracoastal Waterway Systems. The city is served by both freight and passenger rail lines, and has access to Chicago's Midway (35 miles) and O'Hare International Airports (40 miles). Thus, Joliet, like Baltimore, has the potential to be a major distribution center.

Joliet is a home-rule city that adopted its current council-manager form of government in 1955. However, the transition to a "textbook" council-manager form was never fully realized. Consequently, the best outcomes of adopting that form—for example, a professionally run, nonpolitical bureaucracy reporting only to the city manager, and personnel policies based on a merit system—which promote efficiency and a business-like atmosphere were not realized. Council members are elected both from districts and at-large; the mayor is popularly elected. Elections are technically nonpartisan, although candidates are typically classified as "pro-union," "real estate developers," or "bankers." The city manager serves at the pleasure of the council. There have been several attempts to limit the authority of the manager or increase the power of the technically weak mayor in the form of proposed referenda to abolish home rule or the council-manager form.

Former Illinois Governor Dan Walker described the existence of the abundance of governments in Illinois as an opportunity for increased intergovernmental cooperation among units in the state (State of Illinois, 1976: i). Others have argued that increased fragmentation leads to confusion, deflection of responsibility, inequality in service distribution and inefficient government operations (e.g., Neiman, 1982; Hill, 1974; Rehfuss, 1976). The city of Joliet contains seven separate taxing districts—the city, the county, two grade school districts, a high school district, the township, and the port authority. In addition, seven regional organizations are involved in economic development and industrial retention and attraction activities—Joliet Economic Development Commission, Will County Economic Affairs Commission, Greater Joliet, Inc., Downtown Redevelopment Council, Will County Local Development Corporation, DesPlaines River Valley Enterprise Zone Organization, and the Chamber of Commerce. It appears that fragmentation in the Joliet area has led to deflection of responsibility and a situation in which no actor can provide all necessary economic development information. Unlike Baltimore, the Joliet area has no "one-stop shopping center" for individuals or industry interested in development.

The 1980 population of the city of Joliet was 77,956—this was a decrease of over 1,000 from its 1970 mark. Will County (Joliet is the county seat) has experienced a population increase—from 247,825 in 1970 to 324,460 in 1980. A large portion of the area's residents are first or second generation immigrants of the "third wave"— predominantly from Eastern Europe and Ireland. These residents provided a large pool of unskilled and semiskilled labor for capital and durable goods production.

Joliet is an aging industrial "blue-collar" community (Kornblum, 1974). In the early 1900s, food distribution (on the railway system) and steel production were the area's major industries. By the 1960s, steel production declined while the petrochemical industry became dominant. Durable goods manufacturing also figured greatly in the employment picture; for example, two large Caterpillar plants are located in the area. However, this industrial base was not diversified. By 1980, the exodus of a federal arsenal and its suppliers, a steel plant, and a petrochemical plant, combined with a decrease in the demand for farm equipment and other durable goods, put strains on the area's ability to support blue-collar employment. Plants within the city were physically deteriorated while new industrial development was occurring in the suburbs. By 1982, the major employers in the city were governmental or quasigovernmental: the state of Illinois through the correctional system became the largest employer, followed by the two hospitals, the county, city, and school districts. Joliet, thus, had evolved from an industrial and transportation center to a service center dominated by government and a bedroom community for people who work in Chicago and the developing parts of the suburbs.

As with many older industrial cities of the Northeast and Midwest, Joliet has experienced marked deterioration of its infrastructure and its downtown area. An increased inmigration of blacks and Hispanics into the east side of the city resulted in white flight to the west. Poor service delivery to the older areas of the east side contributed to slum-like conditions in and around the downtown area. Joliet reacted to business flight from the downtown and white flight to the west by frantic annexation, actually doubling its area between 1960 and 1971. By the mid-1970s, the only discernable economic development strategy seemed to be one of chasing the tax base in a patchwork fashion to the west and leaving the downtown and east side for lost.

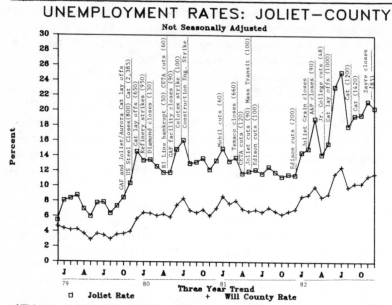

Figure 12.1   Comparison of Joliet and Will County* Unemployment Rates (December 1978-December 1982), Including a Chronology of Critical Events**

SOME CAUSES OF THE HIGH UNEMPLOYMENT RATE AND STAGNATION OF ECONOMIC DEVELOPMENT

Was Joliet's July, 1982 unemployment figure caused by local factors or was it due to factors largely beyond the control of the city? The answer is both. Unhealthy local conditions were exacerbated by national decisions and trends; but the gravity of the economic situation caught most, although not all, officials by surprise. Figure 12.1 documents the trend in city and county unemployment over the period of 1979-1982, and a chronological inventory of critical events that contributed to Joliet's economic situation. This section will outline the external, internal, structural, and political forces that shaped Joliet's destiny.

EXOGENOUS FACTORS

All actors in Joliet agree that the recession and skyrocketing interest rates translated directly into loss of jobs. The construction industry, which in 1970 employed 1,827 city residents and a total of 6,472 county residents, was directly affected as new home starts ground to a halt and construction, in general, declined. The sales of large-scale durable goods, like those produced at Caterpillar, likewise suffered from the effects of high interest rates as farmers increasingly held on to older equipment, making due until a more favorable purchasing climate emerged.

In November, 1980 Caterpillar proposed a large-scale sale to the Soviet Union to compensate for domestic losses. Management anticipated hiring an additional 2,000 people during each of the 5 years required to meet the Soviet pipeline order. However, this hope was crushed on December, 1981 when President Carter blocked all U.S. sales to the Soviet Union.

In addition, the local Caterpillar workers, mostly members of the International Association of Machinists and Aerospace Workers Union, were affected by a strike by the United Auto Workers Union— the union to which other Caterpillar workers in plants in other cities belong. Assembly was drastically cut as supplies from other plants were not forthcoming and orders for Joliet's parts were put on hold. In all, employment was cut from a high of 7,000 workers to 4,000; then, finally, to a point at which less than 600 people—mostly management—were employed at the two plants.

All cities, especially those in the Frostbelt, were affected by the interest rate structure. McGouldrick and Petersen (1968) contend that small local governments are the most adversely affected by rapid increases in interest rates. However, those cities whose economies depended on the industries most affected by inflation and high interest rates suffered even more. Joliet's reliance on the manufacture of durable goods and the supplying of both materials and labor for construction put it in a no-win spiral that escalated the erosion of the area's economic base. Only a diversified economy could have taken up some of the slack and mediated economic hardship. However, Joliet was not alone in this situation; other factors must have contributed to its position.

PLANT CLOSURES

Industrial plants were traditionally drawn to Joliet because of the good supply of land, labor, and raw materials, combined with access to rail, land, and water transportation. Historically, though, the city of Joliet has been, if not unfriendly, at least indifferent to business needs. This attitude can be traced as far back as 1906 when Corn Products Refining, the company that produces Argo Corn Starch, first indicated an interest in locating in Joliet. Apparently, the citizens did not want any more "outside" industry coming into the city. The company gave up the idea of locating in Joliet choosing, instead, to build in an unincorporated area that was to become Argo, Illinois. It was not until World War II, when the federal arsenal opened, that the city realized that industrial expansion would be advantageous to its economy.

The current round of plant closings began in the late 1960s when the federal government closed the arsenal, consolidating it with its Rock Island, Illinois arsenal. Although some of the arsenal employees migrated with the facility, the vast majority lost their jobs. Of those who moved to Rock Island, many returned to Joliet within a year. In addition, Joliet area companies that supplied the arsenal moved to Western Illinois leaving many employees behind.

Between late 1979 to 1982, at least eight companies closed their doors or pulled out of the area with a net loss of at least 2,400 jobs.[2] In addition, permanent layoffs of at least 850 more were reported.[3] Could Joliet have done more to retain these industries or protect jobs? There is evidence to suggest so.

U.S. Steel operated a rod mill near Joliet's downtown area close to both the water and rail lines. With the national cutbacks in steel production in the 1970s, the Joliet plant experienced a slowdown; first temporarily, then permanently, laying off employees. The plant closed in late 1979. The Joliet plant, being relatively old, had a deteriorating physical plant. City sewer and water services to the area east of downtown (the location of U.S. Steel) were also in bad repair as the city concentrated its efforts on providing new service to the west side. The city did, however, confront the steel plant concerning the level of pollutants it was discharging into the canal. In other words, the city was not providing an equitable level of service (based on existing need for improved services) to the plant, but was quick to point out the plant's failures. Having adopted a retrenchment strategy, the plant was not in a position to absorb increased expenditures and was not particularly pleased with the ongoing intense conflict with the city. In the end, U.S. Steel allowed its plant to deteriorate to such an extent that it could justify its pull-out by demonstrating that the Joliet plant was its least profitable plant. That made the decision to close politically neutral—profitability rather than the ongoing political conflicts could be cited for the decision.

In the 1940s and 1950s, a job at the Texaco refinery in adjoining Lockport was considered prestigious. Pay raises remained commensurate with the growth of the petroleum industry. In the 1970s, Lockport and Joliet officials constantly complained to Texaco that the fumes from its plant were noxious and that it should do something to alleviate the problem. At the same time, Texaco employees were demanding higher salaries while the company was complaining that productivity was decreasing. This situation provided a triple financial blow to the company: (1) more money was needed for pollution abatement; (2) more money was needed to provide higher salary packages for employees; (3) more money was being lost due to decreased productivity. In January, 1980, an industrywide refinery strike affected 500 Joliet-Lockport Texaco employees. Texaco bargained separately with its employees and the strike ended April. In October, 1980, the Joliet-Lockport Texaco Refinery closed its doors. The Lockport operation was judged unprofitable, the plant, economically obsolete.

Two factors contributing to the Texaco and U.S. Steel closures were the hostile relations between industry and local government and low employee productivity. In response, both companies allowed their physical plants to deteriorate. In case of U.S. Steel, this infrastructure decay went virtually unnoticed. At Texaco, however, the decay was *visible* as the plant was literally falling apart in full view of area residents. It was clear that the company was not investing in maintenance and it was not until the Texaco closure was imminent that Joliet and Lockport officials asked the company to reconsider its decision. By that time it was too little, too late—it was too expensive at that point for the plant to reverse the trend and continue operations.

Compare this situation with Baltimore's. Baltimore's job pool is known to be highly skilled and productive. Productivity did not appear to be a high priority with Joliet workers or at least with their union bargainers. Workers and unions became complacement during the boom years bargaining for bigger and bigger wage packages without heeding industry's warnings regarding decreasing productivuty. In terms of local governments' relations with industries, those in Joliet can be characterized as antagonistic rather than facilitating; when local governments *did* offer assistance, it was inadaquate. This antagonistic behavior is in sharp contrast to Baltimore's aggressive policy—providing continual contact and support for existing industry. In that atmosphere, contact becomes routine rather than crisis oriented; potential problems can be discussed rather than current practices attacked. The causes of the antagonistic relations between the city of Joliet and industry are explored below.

INSTITUTIONAL INERTIA

Joliet suffers from institutional inertia. Those who hold political power suffer from it; those who control economic resources thrive on it; the citizenry wallow in it; and few people care about it. The root of the problem seems to be a conservative parochialism that pervades the Joliet lifestyle and consciousness.

Joliet is a conservative, ethnic city. Being predominantly third-wave migrants, nearly everyone knows someone or has a relative who is from the "old country." Those immigrants came to America with strong ethnic and religious (mostly Catholic) heritages but few skills or material resources. By working hard, saving their money, and helping those of their own kind, these immigrants built ethnic neighborhoods around their churches, the aggregation of which was to become the city of Joliet. These are a proud people who paid their debts and never asked for "outside" help.

The parochial conservatism has dominated the political, economic, and social structures of the city. Jolietans will, for example, do practically anything to insulate themselves from "Chicago" and all that the city connotes. In fact, it was considered a major political coup when the mayor was able to have the city declared a primary statistical area *independent from* the Chicago metropolitan area.[4] This is also demonstrated in current economic development efforts to draw new industry to Joliet. Rather than capitalize on the low cost of the land and space relative to Chicago's and provide warehousing or bulk processing services for Chicago industries (like many of the not-so-well-situated suburbs are), the city politicians hope to compete with Chicago and the rest of the suburbs for capital and durable goods producers and high-technology firms (for which there exists no available local labor pool). The city staff, on the other hand, is more inclined to develop linkage with Chicago. Unfortunately for the city, this is a case where politically expedient attitudes override professional common sense.

This parochialism is part and parcel of how politics is conducted in the city. Most city business has been conducted with the "electoral connection" in the minds of elected officials.[5] Resources are typically allocated on the basis of benefits conferred to elected officials' districts; parity—determined by counting projects—among politically important districts is an inviolable rule. Under this decision rule, resources are allocated according to political clout rather than need. Unfortunately, areas with the most need tended to be in older parts of the city inhabited by politically silent blacks and Hispanics. In addition, the emphasis on spending money on *visible* projects for which the council member could take credit led to neglect of the city's infrastructure. By the mid-1970s, the Environmental Protection Agency forced the city to separate its storm water and sewage systems or risk the agency's withdrawal of funds to construct west-side sewage treatment plants.

Not surprisingly, political conservatism extends to financing local government. For years Joliet hid deficits in its budget, attempting to keep taxes low by underfinancing pension funds and using nonaccrual accounting methods. Researchers and citizens alike suggested that this was done under the facade of keeping visible service levels consistent in politically important parts of the city without raising taxes (the electoral connection, again). When the citizens learned the extent and duration of what had turned into a fiscal crisis, politicians decried the crisis and deflected responsibility from themselves. The council blamed the city staff; staff blamed the council; "politically" responsive staff blamed the manager. However, even during an era of fiscal crisis, the city had to be literally *forced* by the HUD area office to apply for their Community Development Block Grant (CDBG) entitlement funds. And when they did, it appeared that HUD priorities and legal restrictions were ignored. The HUD area office prodded the city to rewrite the application being more sensitive to CDBG eligible funding priorities. In a sense, Joliet was saying, "If we have to take this federal money, we'll spend it as we like." As late as April 28, 1982, this hostility over governmental grants was summed up by the mayor when he declared that tax money would be better spent if cities like Joliet did not grab up all their available (entitlement) money since some (federally funded) improvements in Joliet were really unnecessary. He went on to assert that rather than deal with imposed national priorities, the city should concentrate its efforts on its major problem: not the prior month's 19 percent unemployment rate, but the illegal conversion of single family homes to multifamily dwellings in a middle-class neighborhood![6]

Again, contrast this behavior with Baltimore's aggressive strategy of combining financial packages from many sources and leveraging private sector resources; how Baltimore used CDBG funds for economic development activities; how it received not only funding but fame through its demonstration grants. Joliet cannot have it both ways. It cannot be isolated from Chicago's influence, ignore outside funding sources, and still compete in a national and international marketplace.

The inertia characterizing Joliet's political system is further evidenced by the behavior of the city council, which is highly reluctant to delegate any responsibility. Rather than dictating policy and allowing the professional administrators to implement policy, council members—and increasingly the mayor—are involved in almost every programmatic decision in the city. This behavior defeats the purpose of having a full-time manager and a part-time mayor.[7] Why does this occur? Several interviewees observed that the fragmented governments in the Joliet area historically have been power stations of family and special interest group "dynasties." One economic observer went so far as to say that he could see no rationale on which decisions were made in the council except personal gain. While this may be an overstatement, the following example provided by an interviewee is sufficient to illustrate the point.

A New York based manufacturing company decided to locate a facility on an undeveloped site on Joliet's west side. When the local representative applied for a city building permit, he was advised by a city employee that the required paperwork was not "in order" and that there was nothing the city employee could do to alter that situation. The company representative could not understand why the city employee would not attempt to untangle the problem or at least suggest the name of someone who could. The local representative responded by calling the corporate office with the recommendation that they abandon plans to develop in Joliet. The home office representative, in turn, called the Joliet real estate developer involved and asked if his firm's experience was typical of standard local practices; he then announced the firm's intention to relocate elsewhere. The real estate developer had to make several calls to various city officials to rectify the situation.

When this scenario was described to a downtown financial executive, he knew exactly what happened. The "wrong" real estate developer signed the papers. When asked what he meant by that, the executive said that all real estate deals must go through the "right councilman" before the permit could be "cleared," that is, placed on the list of permits that can be issued. While one may speculate about the council member's motives (i.e., whether personal gain was operative), it is more important to extract from the situation the instances of institutional inertia: (1) decisions are made politically, under conditions of uncertainty, and are not evenly applied; (2) no apparent institutional apparatus exists to handle nonroutine transactions; (3) city employees, for whatever reason, do not assist businesses to develop or expand.

A traditional risk-avoidance behavior also characterizes the role of the financial community. One bank vice-president gave this impression of the bank's role: "The banks are the best citizens in town" and "A bank is as good as the community it exists in." In Joliet, a good citizen is one who sits on a number of community boards, according to this executive. This bank vice-president volunteered that the downtown banks are interested in downtown redevelopment because all vice-presidents of his bank were expected to sit on community boards; in addition, banks contribute money to agencies like the Downtown Redevelopment Council. However, when asked whether he would be willing to provide a business loan to a coalition of prominent citizens intending to rehabilitate an existing structure for retail and light repair space, he replied that that action would require a counter-sign of personal notes from all the investors. He would not take risks, even with known and presumably dependable entities. A savings and loan officer assumed that a bank auditor would find that type of loan a questionable practice even though another banker actually required this personal lien. If banks are reluctant to make business loans to groups of prominent citizens, what is the likelihood they would provide venture capital to a relatively unknown investor?[8] Since bankers are normally conservative, they can have no real effect on economic development—their reluctance to take a risk lends inertia to new enterprise.

The inertia and conservative ethos found in the political and financial sectors if reinforced by the social structure of the city. The demographic makeup of the city is changing with an increase in its minority and elderly populations. Minorities in Joliet have not been politically active at the city level, save for racial riots after the assassination of Dr. Martin Luther King, Jr. This is partially due to the fact that that the city of Joliet does not provide the types of services that new migrants, especially those with lower incomes, demand: general relief, low-income housing, and health care. To the extent that these services are provided in the Joliet area, they are provided by the county and township. Minorities may believe that they would waste their political energy on an entity—the city—which provides virtually no rewards. The role of private social service agencies in Joliet has increased in importance in recent years, but its significance in no way compares with that in comparable areas of Illinois (e.g., Rock Island County). With the flight of younger middle-class families to the suburbs and the lack of inmigration of like people to the city, the politically relevant public that remains is an older, ethnic population; a traditionally conservative population. This cultural conservatism reinforces the behavior of the conservative political and financial institutions in the city.

Not only are the city officials relatively isolated from the public, but Joliet's citizens exhibit a low level of political efficacy and political trust. Interviewees frequently volunteered opinions such as "city politicians are corrupt," "it does not matter who you vote for—they're all the same," "what I think really does not matter; things will just go on as they always have." They tend to associate a governmental position with an individual or family that has occupied the position, viewing the official's role as that of a caretaker who assumes no importance until the citizen "needs" something. This was demonstrated at one of the mayor's district meetings. Citizens were asked to provide feedback or direction for city staff and council representatives. Virtually all "opinions" were in the form of demands: "Why haven't you fixed the potholes on the 1200 block of Northwest Street?" "What are you going to do about the abandoned home on Cora Street?"[9] Thus, citizen participation takes a reactive form rather than a facilitative directional form. This parallels the way city officials have dealt with industry.

In sum, the conservative, parochial ethos in Joliet permeates its political, economic, and social structures. Such pervasive and reinforcing attitudes shape the behavior of all actors resulting in institutional inertia affecting decision making and the possibilities for economic growth and change in the city.

SUMMARY

There were a number of factors that contributed to Joliet's economic stagnation. The effects of high interest rates and the recession were most keenly felt in cities with nondiversified economies, especially those concentrated in the manufacture of capital and durable goods. Joliet fits this description; but so do many others that have fared better—Milwaukee, for example. However, Joliet was plagued by other exogenous factors: national strikes that affected nonallied workers at the Caterpillar plants and the Carter decisions restricting foreign trade.

Joliet was the victim of a series of plant closures between 1978 and 1982 resulting in a net loss of at least 2,400 jobs. The decisions to close some of these plants were directly linked, at least in part, to decreased worker productivity, deteriorating physical plants, and continual friction between the industries and the city. The role of the city with regard to relations with industry was reactive and crisis oriented rather than ongoing and negotiating in nature. Consequently, public and private officials lacked both a forum to "clear the air" of potential problems and also a nonthreatening arena to get to know each other. A friendlier business climate might have prevented some plant closures.

Joliet suffered from inertia and parochialism in its political, economic, and social institutions. Even routine functions in government were operated under conditions of uncertainty; they were handled in a political arena. The result was inconsistent responses to industrial inquiries and an overall distrust of local government by both industry and citizens alike. City officials strove to isolate the city from Chicago and the influence of "strings" attached to federal government support. In doing so, they bypassed opportunities to recruit Chicago industries to take advantage of the low-cost space available for warehousing and bulk processing. By not aggressively taking advantage of available government grants, they denied the citizenry its share of federal tax subsidies. Without such monies, the city did not have the capital to engage in economic development negotiations with industry at a level competitive with other cities.

The Joliet financial community has only paid lip service to participation in economic development activities. Being in the business of making money for their depositors, they are not overly anxious to incur risk. The conservative atmosphere in the city and the uncertain economic climate has exacerbated this behavior. The climate to support business loans or engage in venture enterprises is absent.

This behavior on the part of the political and economic sectors is reinforced by a conservative and parochial ethos prevalent in the city. Dominated by older ethnic citizens, the city does not operate in an atmosphere conducive to change. Political efficacy and political trust are both at a low ebb. The citizens perceive, and rightly so, that the political processes in the area have been dominated by family and special interest groups. Unlike the domination of, for instance, the Ball family in Muncie, Indiana or the Mellon family in Pittsburgh, no one family or set of actors has the stature, influence, or will (or, maybe even the money) to really affect comprehensive change for the betterment of the city. Given all these factors, are the cards stacked against any economic development efforts in or around the city of Joliet? The next section describes some economic development efforts attempted by Joliet while the last provides a prognosis for the area.

THE STATE OF ECONOMIC DEVELOPMENT IN JOLIET

By the early 1970s, Joliet's downtown was virtually dead. Many stores left the downtown area to join the two new major shopping centers: one on the west side and one in a newly annexed section of the city far to the northwest. Included in these moves were J. C. Penney, Sears and Roebuck, Woolworth's, and a number of other jewelry and clothing stores. By the 1980s, even the main post office moved west. What remained downtown were the new country courthouse, new city hall, four major financial institutions, older office buildings, and a few retail and restaurant hangers-on. Two relatively new office buildings were constructed by that time and three of the financial institutions engaged in large-scale renovations and reconstruction. However, only downtown workers and students from a nearby high school frequented the area.

The Downtown Redevelopment Council was organized to revitalize the downtown area and bring life back to the east side. Its initial efforts were directed at constructing a downtown open-air mall, providing off-street parking, providing new office space for people who needed speedy access to the city and county complexes, and renovating the Rialto Theatre and Rialto Square buildings.

The mall and parking deck projects sparked heated controversy in the city council, which took a number of *years* to iron out. The details of the controversy are colorful (and even cost some council members reelection); however, most of the problems encountered could have been predicted given the city's history. First, all matters concerning bidding on projects, design, allocation of parking slots, and bond floating, to name a few, were decided in a political forum—one that has been characterized as conservative, risk averse, and highly volatile. By the time the projects were completed, still more merchants had left the downtown area and the anticipated costs of project completion skyrocketed. Second, a whole series of maneuvers concerning the assessed devaluation of downtown property and an increase in valuation of single-family homes in tandem with an increase in the property tax left the citizens in an uproar. The number of contested tax assessments during that period only exacerbated Joliet's financial problems and indicated the extent to which the citizens were displeased with governmental operations.[10]

However, the mall and parking decks are now in place and the Downtown Redevelopment Council is intensifying efforts to draw workers and shoppers back to the downtown. The first priority is to draw service workers (such as lawyers) from the west side by providing rehabilitated and new office space at extraordinarily low rates. The council has also sponsored a series of festivals on the new mall to reacquaint the citizens with the downtown. Once the worker pool in the downtown area expands, the council hopes to draw new merchants and restaurants to the area, which would, in turn, draw shoppers and new life to the area. The success of these plans has been affected adversely, however, by the unsolved daytime murder of a prominent citizen in one of the parking decks. Many citizens are afraid to go downtown.

The renovation of the Rialto Theatre, a beautiful center dating from the days of burlesque, is also considered to be a boom for economic development in the downtown. In May 1977, the Rialto, then a movie theater, was in danger of being demolished by its tax delinquent owners. An effort spearheaded by historic preservationists resulted in an eventual bail-out by the Illinois State Legislature, which established an independent district to oversee the renovation. In the process, though, the city and the owners were involved in an ongoing political controversy concerning the theater's fate. Had it not been for the state bail-out, the theater would have been demolished as the debate in the city was framed in terms of a controversy rather than by negotiation. In the end, Joliet has a beautifully refurbished theater for which the city of Joliet provided *no* monetary support.

Future plans include the renovation of the historic Union Station to include, among other things, a first-class restaurant to complement the activities at the Rialto. Office buldings continue to undergo renovation. Although an all-out effort has not been undertaken to draw retail establishments back to the central district, the Downtown Redevelopment Council believes that the retail establishments will come back as they see the growth in downtown worker population. Finally, the city and number of neighboring communities have been awarded an Illinois Enterprise Zone status by the state in the hopes of attracting small industries and tourism to the Illinois and Michigan Canal Industrial Valley. This area currently has a large number of abandoned buildings. The zone covers a much larger physical area and number of legal participants than that suggested by the federal design, which could conceivably cause future problems. Moreover, it is not clear that the enterprise zone will have much effect as the benefits conferred by the State Enterprise Zone law are actually less than those available by taking advantage of current federal tax laws. This situation is true in both the state and proposed federal legislation. This appears, however, to be a first step toward acknowledging development activities.

JOLIET'S PROGNOSIS

DOES JOLIET HAVE THE CAPACITY TO ATTRACT NEW INDUSTRY?

All of the organizations in the area currently engaged in economic development activities are officially concerned with retention of existing industries and attracting new ones. Representatives of Greater Joliet, Inc. recently traveled to Switzerland for a conference organized to afford American communities the opportunity to "sell" themselves to Swiss industry. Although this is activity aimed at expanding Joliet's economic base, it may be more appropriate for the city (and area) to exploit (1) areal relationship with Chicago, (2) excellent transportation access, (3) existing structures, and (4) available labor force to quickly alleviate the current economic hardship. It is hoped that these can become long-term activities.

Unfortunately, Joliet has many problems that act to discourage executives from locating industries within its boundaries. First, the city is physically old and dirty. With its aging population, the number of school children (and young executives) is rapidly declining, and along with it the public school system. Although the Rialto renovation is a first step, there are no other major cultural ammenities in the immediate area. There are no first-class restaurants. All in all, Joliet is not an attractive place to locate.

The unpredictable and fragmented political system is not conducive to conducting business. The chamber of commerce, the organization one would think would be a prime contact for interested executives, is ineffective. No organization involved in economic development activities can provide the services of, for instance, Baltimore's BEDCO.

The parochial atmosphere that pervades the city has to become more flexible if new companies are to become a part of the area. The city has to assume an areal view of itself in relation to Chicago. It might well key-off its proximity to Chicago and provide services that are not available there, such as inexpensive rental space for warehousing and bulk processing. In the words of one interviewee, "Joliet has to decide whether it is going to go ahead now or sink into oblivion."

ARE THERE ANY BRIGHT SPOTS IN JOLIET'S FUTURE?

In October, 1981, Joliet hired Kenneth R. Murray, former Deputy City Manager of Grand Rapids, Michigan, as the new city manager. Murray had participated in Grand Rapids' downtown development program and was thus seen as a person who could lend some credibility to Joliet's downtown effort—the only development program that was receiving any popular support. Early in his term, Murray placed a premium on hiring trained urban planners and community development specialists, hired a full-time grantsman, made concerted efforts to disentangle political considerations from the city's administration and personnel, and made economic development the number one priority of his administration.

However, having qualified personnel in place does not ensure either that program changes will be implemented or that such efforts will not be hampered by the same parochial attitudes that pervade the political system and citizenry at large.[11] Murray and his staff are aware of the basic conservative distrust of all governmental activities in Joliet. Therefore, they have attempted to develop a grass-roots awareness plan at the council district level. They hope to promote meaningful citizen participation in the city's planning and to increase awareness of alternatives for allowing the city to grow again. For instance, at district meetings, Murray and his staff divide citizens into small groups with one city staffer. They utilize nominal group technique to devise lists of *community* concerns (rather than personal complaints). The concerns are then discussed in the committee of the whole. The citizens develop a community strategy that identifies the type of community they wish to live in and the actions they are prepared to underwrite in order for that future to occur.

City staff are also involved in courting industry and assisting business executives in their dealings with the various political units and economic development agencies. The manager and staff attempt to "cut the whole deal" with the interested party *before* they have to meet the political, financial, and real estate interests. In the interviews, it was clear that the city staff did not trust the citizen leaders to make prospective industrialists and executives comfortable with city operations and development opportunities.

Attempts like these are important steps toward developing a sense of trust in government that would probably lead to a more stable consistent political climate in Joliet in the future. However, radical change cannot be expected in the near future as Joliet requires a major overhaul of its political, social, and business institutions and an abrupt change in the attitudes of its citizens to compete with other cities in the region. Given regional trends, Joliet will be swimming upstream. One can appreciate how far the city still has to go to achieve economic health when one considers, for example, that the county board decided to delete the economic development line of their CDBG application with the rationale that they could engage in those sorts of activities when the recovery arrives.

WHAT HAVE WE LEARNED FROM THE JOLIET EXPERIENCE?

It is not enough to say that Joliet's experience with economic development appears to be disastrous without attempting to pull together the various lessons that other cities can learn from Joliet. Although Joliet's experience cannot be generalized to all middle-sized cities in the industrial Northeast and Midwest, it, unfortunately, has encountered many barriers to successful economic development. Middle-sized cities may identify with one or more of Joliet's prob-

TABLE 12.1   Characteristics and Strategies that Affect Economic Development: A Comparison of Joliet and Baltimore

Characteristics in Common

Land available for development
Good transportation access
Available labor pool
Capable city administrator

Characteristics at Variance

(What Joliet lacks and Baltimore has)

Characteristics of the City	Capacity of City Government
Skilled, productive work pool	Stable, consistent political environment
Public trust in government	Ability to spend money on economic development (good credit rating)
Support for economic development activities	Assume dollar risks for economic development
Solid infrastructure to support new development	Ongoing dialogue between government and industry
Attractive appearance	Risk to undertake new programs
Cultural amenities	Expertise to creatively develop funding packages for leveraging
	Political structure dedicated to economic development activities

Development Strategies

Areawide development strategy
One-stop center for development information
Comprehensive downtown development strategy
(office, retail, housing)
Vocational and manpower training component for long-term strategy

lems, acknowledge how these affect successful development, and take positive action to deal with their respective problems. Table 12.1 provides a list of characteristics and strategies cited in this and the previous chapter that affect the success of economic development.

While Baltimore's success could be considered by some as an aberrant situation due perhaps to Mayor Schaeffer's "pull" in Washington, this case study lends some evidence to suggest that cities with some development potential (areal, transportational, available labor pool), but which lack all the positive attributes exploited by Baltimore, are in serious trouble. In other words, from the absence of these characteristics one can predict economic failures. This does not mean that Baltimore "wrote the book" on economic development. Rather, Baltimore has exploited available means on a grand scale even when ventures were considered risky. The city of Baltimore took responsibility for its future.

The Committee for Economic Development (1982) has reiterated this responsibility theme. They argue that since economic conditions are uncertain, local governments can help counter that uncertainty by providing an environment of predictability in which developers can operate. Local governments can provide the arena in which differences can be mediated and ongoing communications networks established. They can also act as a conduit for the incubation of civic entrepreneurs—persons with leadership ability and the ability to increase public confidence and develop consensus on development goals.

Richard Moore and Michael Pagano (1983) have found that when small local governments are adversely affected by rapid increases in interest rates, they abandon capital improvement programs—usually to continue allocations to more visible services. It is incumbent for such city councils to be farsighted enough to maintain infrastructure to support local businesses and provide for potential new development.

Gail Schwartz (1983) reports that local governments should also take positive action to alleviate what she describes as "market friction"—inefficiencies that occur in identifying capital sources, linking unemployed workers with available jobs, supplying producers of goods and services with information regarding consumer demand, and reducing government regulation. In addition, she argues that the public sector must try to reduce the cost of production for business through such devices as tax incentives, encouraging low-cost loan programs, and leveraging funds for development. Here, local government can encourage financial institutions to reshape their loan policies to meet the needs of old and new developers alike. This would require some flexibility on the part of financial institutions and perhaps increased bank risks that could be underwritten through any number of private and public partnerships. The American Bankers Association (1982) stresses that effective private participation in re-development efforts depends on several critical variables including the commitment and activity of the financial institution's CEO, the diversity of bank approaches as a reflection of different community and bank characteristics, and the commitment and effectiveness of the financial community's other partners—business, government, in-dividuals, and community-based organizations.

Unfortunately, it appears that Joliet had neither the capacity nor the commitment to engage in meaningful, comprehensive develop-ment activities. The inertia found in all sectors precludes progress at more than a superficial level. Recently, however, Joliet, through the city manager's office, has initiated some development activities and even gained political support for some form of long-range planning. But continued political support is not guaranteed. The attrition rate among Joliet's city managers is high—the seven have served an average of three and one-half years, all apparently succumbing to the vascillating volatile political climate.

What we have learned from the Joliet experience? Economic development requires commitment, climate, and comprehension—Joliet has had none of these requisites. In February, 1983, the city reached its highest unemployment rate in history—26.6 percent; again, the nation's highest.

NOTES

1. For an excellent example of a case study of a middle-sized city and the extent to which it exhibits unique problems—not just smaller-scale problems of large cities—see Rubin (1982).

2. In scanning the Joliet *Herald-News* during that period, I found the following closures: Joliet Grain Company, U.S. Steel, GAF Corporation, Texaco Refinery, Pen-Dixie, Diamond, A&P, and Zayre. The estimated loss of jobs was calculated from these news accounts.

3. Included in this number were layoffs at Commonwealth Edison, Joliet Junior College, Job Service, city of Joliet, Mobil Oil, Celoex Roofing, and Alumax.

4. The impetus for the mayor's effort to become a separate statistical area was his conviction that having unemployment statistics listed separately from the metro-politan area's would focus additional state and national attention on Joliet—to make others aware of the city's economic plight. However, the treatment of the event in the media underscored this "separatist" attitude.

5. In fact, one candidate for council proclaimed that his goals were to "get reelected and keep taxes down" (Joliet *Herald-News,* February 6, 1979: 1).

6. The mayor's attitude on this has changed rather dramatically in recent months; he is encouraging staff to seek more outside funding. Attitudes about how to spend the money have not changed, though. When the city received $422,000 in Community Development Block Grant jobs money in Spring, 1983, staff engaged in accounting reallocations that technically met the federal guidelines but had the net effect of distributing the money through the city for councilmembers' district projects.

7. For example, getting a city engineer to consider a less costly sewer and water hook-up alternative (let alone provide a site inspection) for a city resident required the intervention of a councilman.

8. In fact, another small firm left the city since it could not secure a $35,000 rehabilitation loan from any Joliet bank.

9. The mayor was pleased with the turn out and number of "suggestions" from the citizens. This seems curious as demands do not provide direction. City staff, realizing this, would not participate in these public flogging scenes and, rather, changed the format of the meetings in an attempt to gain a clearer view of where citizens felt the city should be going.

10. Tax assessments and valuations are made by the county. However, the timing of the devaluations and tax increases left even the most casual observer concerned about possible collusion among the county, city, and downtown real estate owners.

11. In fact, although Murray's appointment was lauded in the press as a move toward a more professionalized city bureaucracy, it is commonly held that he was chosen over the "local" candidate (the then acting city manager) because of a political faux pas—the "local" candidate surprised the council with a proposal for a refuse collection tax immediately after they had expended considerable effort to cut taxes and services.

REFERENCES

American Bankers Association (1982) "Commercial banks," pp. 11-34 in R. A. Berger, K. S. Moy, N. R. Peirce, and C. Steinbach (eds.) Investing in America: Initiatives for Community and Economic Development. Washington, DC: President's Task Force on Private Sector Initiatives.

Committee for Economic Development, Research and Policy Committee (1982) Public-Private Partnership: An Opportunity for Urban Communities. New York: Committee for Economic Development.

HILL, R. C. (1974) "Separate and unequal: government inequality in the metropolis." American Political Science Review 68 (December): 1557-1568.

KORNBLUM, W. (1974) Blue Collar Community. Chicago: University of Chicago Press.

McGOULDRICK, P. F. and J. E. PETERSEN (1968) "Monetary restraint and borrowing and capital spending by large state and local governments in 1966." Federal Reserve Bulletin 54 (July): 552-581.

MOORE, R. J. and M. A. PAGANO (1983) "Capital formation, supply economics, and the public sector." Southern Review of Public Administration 6 (Winter): 450-464.

NEIMAN, M. (1982) "An exploration into class clustering and local government inequality," pp. 219-234 in R. C. Rich (ed.) Analyzing Urban-Service Distributions. Lexington, MA: D. C. Heath.

RUBIN, I. S. (1982) Running in the Red. Albany: State University of New York—Albany Press.

REHFUSS, J. (1976) "Intergovernmental conflict between Illinois suburban park districts and municipalities." Presented at the annual meetings of the Midwest Political Science Association.

SCHWARTZ, G. G. (1983) "The new realities of economic development." Southern Review of Public Administration 6 (Winter): 390-405.

13

Local Policy Discretion
and the Corporate Surplus

BRYAN D. JONES and LYNN W. BACHELOR

☐ BROADLY SPEAKING, city governments pursue two different kinds of public policies: those affecting consumption and those affecting production (Johnston, 1979: Ch. 2). The *politics of consumption* concern the allocation of public goods and services to citizens and the redistribution of income through taxing and spending. The classic issues of urban deprivation, the distribution of services, and the provision of collective goods all fall within the domain of the politics of consumption. The *politics of production* center on the construction of the infrastructure that is necessary to profitable private investment. Providing necessary transportation networks, constructing water and sewerage arrangements for offices and factories (as opposed to residences), and the provision of locational incentives for capital investments all fall within the domain of the politics of production.

It would, of course, be foolhardy to ignore the connections between production and consumption. Cities tax the commerical and industrial property within their boundaries with an eye toward the provision of services and public sector jobs for their citizens. There is a strong belief on the part of city officials that economic growth is beneficial because such growth generates a public surplus. (Molotch, [1976] and Maurer and Christenson, [1982] provide some empirical evidence). Tax revenues from the added economic increment can be used to provide services and jobs for citizens. Collective consumption is, in the minds of city officials, linked to production—although it is unclear whether government policies intended to increase the productivity of the city economy actually generate such a public surplus (Friedland, 1982: Ch. 9).

AUTHORS' NOTE: *An earlier version of this chapter was presented at the 1981 Annual Meeting of the Midwest Political Science Association, Cincinnati, April.*

While at one level production and consumption are inextricably linked, the politics of consumption and the politics of production take place in isolated arenas. They are linked only at a limited number of interstices. Hence these two domains of city politics form separate dimensions of action, even if the two dimensions are themselves related. Andrew Kirby (1982) has forcefully argued that social conflict proceeds along both dimensions, and that the dimensions cross-cut one another. And Roger Friedland (1982: 202) has written of the existence of

> two worlds of local expenditure: one oriented to providing services and public employment for the city's residents, the other to constructing the infrastructure necessary to profitable private investment. These two worlds—of social wage and social investment—[are] structurally segregated.

In this article, we examine the pure politics of production by focusing on the interactions between the city of Detroit and the General Motors Corporation concerning the construction of a new automobile assembly plant in the city. This facility, the infamous "Poletown" plant,[1] involved the destruction of a neighborhood. It was industrial urban renewal in a grand scale: within 18 months of the announcement of the project, 1,500 homes, 144 businesses, two schools, a hospital, 16 churches, and an abandoned reinforced concrete automobile assembly plant whose demolition cost alone was estimated at $12 million were gone, and 3,438 citizens had been relocated. While a substantial amount of policy maneuvering and rancorous conflict took place, the key issue was simply whether and how the city of Detroit could meet the stringent and inflexible demands of General Motors. There was, in a word, no policy discretion; there was only capitulation.

It is easy to see how this situation emerged from a particular political economy dominated by industrial decline in the Northeast and Midwest, by increasingly mobile capital controlled by a limited number of large firms, and by increasingly vigorous competition among states and localities for scarce capital investment (Katznelson, 1976; Cook, 1982; Bluestone, 1983; Long and De Are, 1983). Each of these factors has enhanced the bargaining power of corporations in their dealings with city governments. On the other hand, substantial evidence exists indicating that public policies have scant influence on corporate locational decisions (Schmenner, 1980; Vaughn, 1979; Lund, 1979; Mandell, 1975). Why, then, was the relationship between GM and Detroit one of capitulation?

We argue that this capitulation had three causes: (1) policy and administrative discretion by local governments; (2) the design of economic development policies; and (3) the asymmetry of information that characterizes the negotiations between corporations and city governments over the location of facilities. The corporation knows just what is necessary for a city government to provide in order to attract the facility, while city officials are not privy to that information. Although in most instances state economic development policies are designed to leave city officials substantial discretion in their application to particular cases, because of the asymmetry of information characterizing the negotiations, officials are inclined to grant concessions up to the statutory limit.

The extraction of concessions by a corporation beyond what would be strictly necessary to attract the facility we term the *corporate surplus*. This surplus is the result of a "hidden hand" of large-scale corporate capitalism superimposed on a patchwork of small territorial governments competing among themselves.

Below, we review the major economic development strategies and employed by state and local governments, indicating how they were used in the negotiations between General Motors and Detroit over the location of the Poletown assembly plant. Then we describe the politics surrounding the extraction of the corporate surplus in this particular case. Finally, we analyze the causes and consequences of the corporate surplus.

ECONOMIC DEVELOPMENT STRATEGIES

The formulation of economic development strategies takes place in the context of bitter competition as older industrial cities contend with the booming Sunbelt states and nonmetropolitan areas of their own states for the tax dollars they so desperately need. The state economic development programs stimulated by this competition now offer so many incentives that they have lost much of their pulling power: Some 25 states offer corporate income tax exemptions, 29 offer a tax exemption or moratorium on land or capital improvements, and 31 provide state financing assistance for plant expansion (Peirce and Steinbach, 1980). At best, these and other state policies can provide a slight advantage to one *state* over another—making Michigan more competitive with Oklahoma, for example—but they have seldom been targeted to "distressed" cities, and so are of little value in attracting investment to a city rather than its urban fringes. The outcome of *intrastate* competition is affected by such factors as land costs, transportation access, and local political characteristics (espe-

TABLE 13.1 Public Sector Revenues Sources and Project Costs, Detroit General Motors Assembly Plant

Source	Total Amount ($)	Acquisition	Relocation	Demolition	Roads	Other Site Preparation	Professional Services
HUD Letter of Credit	65,000,000	60,112,400				4,887,600	
HUD Urban Development Action Grant	30,000,000		16,009,000	10,000,000		3,991,000	
HUD Section 108 loan	35,000,000	33,567,000		1,433,000			
Community Development Block Grant (HUD)	8,522,000	400,000	2,450,000	3,000,000			2,672,000
Economic Development Administration	15,000,000			9,300,000		5,700,000	
Urban Mass Transportation Administration	1,364,000	901,000	363,000	100,000			
State Road Funds	32,660,000	4,530,000	1,570,000	700,000	25,335,600	524,400	
State Land Bank	1,425,000						1,425,000
Program Income Interest*	2,400,000	2,400,000					
Program Income**	11,470,000	11,470,000					
Program Income*** Fixture Sale	1,000,000						
Totals	203,841,000	114,380,400	20,392,000	24,533,000	25,335,600	15,103,000	4,097,000

SOURCE: City of Detroit, Community and Economic Development Department, April 9, 1982.
*On funds put in escrow account during property condemnation proceedings.
**From sale of property to G.M., Conrail, etc.
***From sale of fixtures from businesses relocated from project area.

cially access to federal development funds). In the end, corporate investment and plant location decisions may be determined by factors beyond the control of public officials, such as the presence of a network of suppliers, prevailing wage scales, or the degree of unionization of the work force. Nevertheless, public officials desperate for jobs and tax dollars are understandably reluctant to risk losing a potential source of revenue by offering fewer concessions than their everpresent competitors; although they can't afford to offer the tax reductions, they can't afford not to.

Tax reductions, however, may be less significant in their effects on private investment decisions than the various forms of state enabling legislation to facilitate development projects, or the ability of local officials to utilize federal funds to "leverage" corporate investment. These state and federal programs place those levels of government in the role of encouraging a wide range of forms of private-public cooperation at the local level. The form of public-private cooperation in any particular city, then, will be affected by state enabling legislation and access to federal funds, as well as by internal political characteristics (Clarke, 1982).

ECONOMIC DEVELOPMENT PROGRAMS AND THE POLETOWN PROJECT

Vaughan (1979) has listed more than 35 forms of financial assistance, tax incentives, and special services offered by state and local governments to encourage private sector investment. States with the largest number of programs were concentrated in the Northeast/ Midwest, indicating an effort to compensate through public policies for the other competitive disadvantages suffered by these regions. Among the most popular strategies were revenue bond financing (45 states); tax exemptions on raw materials used in manufacturing (44 states); various forms of inventory tax exemptions, state financing aid for plant expansion (29 states); and corporate income tax or property tax reductions.

City officials, in theory, can deploy any combination of the economic development tools necessary to attract investment. In practice, the city often offers a smorgasboard of incentives to corporations, who hold the power of acceptance or rejection. In the case of the Poletown plant, the city of Detroit used all of the following: (1) the provision of property tax abatements; (2) the administration of the project by a public-private economic development corporation; (3) the revision of property acquisition and eminent domain procedures (which expedited project implementation); (4) the securing of federal grants and loans to pay for property acquisition, relocation, and site preparation; and (5) the establishment of tax increment financing to repay federal loans (for which state authorization was required). All were made necessary by the demands presented by GM, although city officials displayed considerable skill and ingenuity in piecing together a package acceptable to GM and in acquiring the funds necessary to finance the city's share of project costs.

TAX INCENTIVES

Reductions in corporate property, income, or inventory taxes are one of the most widely used and controversial financial incentive programs of state and local governments. Although property taxes are collected by local governments, state authorization is required for municipalities to offer reductions, which then must be approved by local officials. Eligibility standards and specific provisions of tax incentive programs vary considerably among states; property tax abatement authority, for example, can be targeted to certain types of cities, while corporate income tax reductions may be limited to certain types of firms, or be dependent upon increases in employment. Sometimes property and income tax reductions are set up as alternatives, as in Louisiana, which prohibits facilities receiving property tax abatements from receiving corporate income tax credits.

Tax incentives result in immediate revenue losses for state and local governments. Proponents claim that these temporary losses are outweighed by long-term gains in revenue from investment that would not have taken place had the tax incentives not been available. It is further argued that higher taxes in the northeastern and midwestern states contributed to the flight of industry and jobs from those states, and that tax reductions are needed to stem this movement. The consequence has been competition among these "Frostbelt" states to offer the most generous tax reductions to new facilities (Vaughan, 1979). This interstate competition also puts corporations in the position of being able to set one state against another by threatening to move to another which has made a more generous offer.

The most serious shortcoming of tax incentives as an economic development strategy is that there is little evidence that they have a significant impact on business investment or plant location decisions. According to the National Council for Urban Economic Development (1981: 36-37), there are two main reasons for this: (a) the federal tax structure, by allowing deductions for state and local taxes, reduces both the burden of those taxes and the amount of the subsidy provided by concessions; and (b) other aspects of a firm's cost structure, such as labor costs, are much greater, and thus more important influences on location decisions.

Michigan Public Act 198 (1974, amended 1978) authorizes local governments to establish industrial development districts, in which property taxes on a new facility are reduced by 50 percent for up to 12 years, and plant rehabilitation districts, in which the property tax assessment of a replacement facility is frozen for up to 12 years at the level of the "obsolete" plant being replaced. At the insistence of GM and the recommendation of Mayor Young, the Detroit City Council granted tax abatement for the new Poletown assembly plant.

ECONOMIC DEVELOPMENT CORPORATIONS

State constitutional restrictions on local government authority in raising taxes, buying and selling property, and negotiating with private firms can create difficulties for local development programs. By authorizing the formation of local economic development corporations (EDCs), which create private-public partnerships, states can ease these restrictions and combine the powers of the public and private sectors. As with much state enabling legislation, that authorizing EDCs is often passed in response to pressures from city officials, and the use of EDCs tends to be concentrated in large cities. Philadelphia pioneered the concept in 1958 and numerous others have followed suit.

Powers granted to EDCs vary from state to state, and may include acquisition and development of land, issuing industrial revenue bonds, borrowing funds, receiving grants and loans, and entering into agreements for the lease or sale of project sites to private corporations. They may also be allowed to offer such development incentives as land banking, tax abatements, and low cost loans. The granting of such broad powers to quasipublic organizations has raised questions of accountability, because, as one report put it, "When private sector persons participate in the formulation and management of public development programs, they are also making decisions about the use of public powers and resources" (Council for Urban Economic Development, 1978: 9) EDCs, however, may simply formalize cooperative relationships between public and private officials. In our Detroit case study, private sector influence was not exercised through the EDC, but rather through the monopoly of technical knowledge relevent to project planning and the threat of moving the project elsewhere.

Nevertheless, the EDC mechanism was crucial to the planning and implementation stages of the Poletown project. Particularly helpful were the Michigan Economic Development Corporation Act's definition of "public purposes"—as including alleviation and prevention of unemployment, retention of local industries, and encouraging industrial expansion—and another provision that allows municipalities to acquire property through eminent domain and transfer it to EDCs, stating that this "taking, transfer, and use shall be considered necessary for public purposes and for the benefit of the public" (P.A. 338, 1974, amended, 1978; sec. 2, sec. 22). The combined effect of these provisions was to allow condemnation of property through eminent domain powers by the city of Detroit for use in the Detroit EDC-sponsored General Motors project; the EDC then sold the cleared site to General Motors for construction of its assembly plant.

Under P.A. 338, the board of directors of an EDC must have at least nine members, of whom no more than three may be city officials or employees (Sec. 4, par. (2)). To counteract this private sector dominance, the act requires participation of public officials in the planning and adoption of development projects. The Detroit City Council, for example, was formally responsible for establishing the boundaries of the project area, conducting public hearings on the project plan, approving or rejecting the plan, and had the option (which it exercised in the case of the General Motors project) of establishing a citizens district council to represent residents and businesses in the project area. Local development departments are also required to review and submit recommendations on EDC project plans; P.A. 338 sets forth in detail the information that must be included in a project plan. In the case of the Detroit General Motors project, staff from the city's Community and Economic Development Department (CEDD) worked closely with EDC staff in preparing financial and relocation plans. Plans for the assembly plant itself and other physical facilities were formulated by General Motors, and GM staff worked closely with CEDD staff during the project.

ENABLING LEGISLATION

In addition to the revisions of eminent domain powers in P.A. 338, changes in property acquisition procedures were essential to the implementation of the Detroit General Motors project. Michigan P.A. 87 (1980), adopted April 8, 1980, referred to as the "quick take law," allows public agencies to secure title to property acquired through eminent domain before reaching settlements on compensation for property owners. Under P.A. 87, the agency must establish "just compensation" for each parcel it seeks to acquire, and submit an offer to the property owner for no less than that amount; if a purchase price cannot be negotiated with the owner, the agency may file a complaint for the acquisition of the property in circuit court, asking the court to determine just compensation. At that time, owners may challenge the necessity of acquisition, but the act prescribes limited conditions for this review, and sets time limits for the filing of complaints, scheduling of hearings, and rendering of decisions on these challenges. If no challenge is filed, or if the challenge is denied, title to the property is turned over to the agency, which must pay the owner the previously determined "just compensation" pending court review of the acquisition price. Prior to the enactment of P.A. 87, public agencies could not obtain title to property *until* an acquisition price had been agreed upon, a condition that delayed and sometimes prevented the implementation of urban renewal projects. Without these new procedures, Detroit could not have met the strict deadlines for site preparation set down by General Motors because it would not have obtained title to all property on the plant site.

FEDERAL FUNDING

Site preparation could not have been completed without funds from the federal government. Detroit's critical fiscal situation, which was responsible for city officials' pursuit of the plant as a means of stabilizing a shrinking tax base, made it impossible to finance site preparation costs of $200 million through bond sales. Major sources of funding were loans and grants from the Department of Housing and Urban Development (HUD) and a grant from the Economic Development Administration (EDA).

Two HUD programs figured importantly in financial plans for the Detroit project. Through the Urban Development Action Grant (UDAG) program, the city of Hamtramck (in which part of the project site is located) received a $30 million grant, which was budgeted for relocation, demolition, and site preparation. This represented the largest grant ever awarded under the program, which was established in 1977 to aid severely distressed cities in alleviating physical and economic deterioration by providing funds to stimulate private investment that would not otherwise have been made.

The UDAG application, however, was not approved until February 1981, and money was needed in the fall of 1980 to cover acquisition and relocation costs. The need for quick money led city officials to explore HUD's Section 108 loan program, under which cities can use Community Development Block Grant funds as collateral for loans for property acquisition or rehabilitation. Loans can total up to three times the city's annual block grant allocation, and block grant funds must be pledged to repay the loan if project revenues are insufficient for repayment. At the time Detroit submitted its application, 24 projects in 23 cities were receiving Section 108 loan funds; Detroit had previously received $38 million from this program for a convention arena and parking garage. The city received a total of $100 million for site preparation for the General Motors project.

In order to save interest cost on the Section 108 loan, the city financed immediate project costs through a "letter of credit," which allowed the city to borrow committed but unspent Community Development Block Grant funds from current and past years without incurring any interest costs and without the delay involved in applying for federal grants. As of October 1980, when the project plan was approved, Detroit had $60 million of that amount in an escrow account to cover project acquisition and relocation costs.

The other main source of federal money was the Economic Development Administration, which provided a $15 million grant. Initial financial plans had projected $30 million, in several installments, from E.D.A., but these plans had to be revised when the Reagan administration cut this program. The difference was covered by the Section 108 loan.

TAX INCREMENT FINANCING

Reductions and anticipated reductions in federal grant programs, heavy reliance on federal loans, and an uncertain economy forced city officials to explore alternative means of repaying project loans. Tax increment financing, which allows the establishment of special development districts in which projected increases in property tax revenues are used to finance additional public improvements or to repay initial financing costs, had already been authorized by state law for central business districts. Lawyers from Detroit's Economic Development Corporation participated in the drafting of legislation extending tax increment financing to development areas outside central business districts, which was adopted in January 1981 (P.A. 450 of 1980). The tax increment financing plan for the Central Industrial Park (General Motors) Project adopted by the Detroit City Council on April 21, 1982 committed more than $5 million in project property tax revenues annually to repayment of project costs for a 30-year period (the amount varies from year to year but stablilizes at $7.6 million per year after the expected expiration of the tax abatement in 1997); only $169,150 would be retained by various units of local government, primarily the city of Detroit and the Detroit Public Schools.

The plan is based on estimated project costs of $203 million, estimated State Equalized Value (SEV) of the assembly plant of $159 million, and estimated initial SEV of property in the project area (as of May 1981) of $2.1 million. Tax increment financing commits tax revenues on the difference between initial SEV and postproject SEV to the repayment of project cost: thus, the tax increment financing authority receives the tax revenues on $157 million and the combined local governments receive the tax revenues on $2 million, until the year 2012. The "initial SEV" used in the tax increment financing plan represents the value of property in the area *after* most demolition had taken place; more than $10 million additional SEV would have been retained by local government had the assessed value of property in the project area before the project was initiated—estimated at $13.3 million in the project's Environmental Impact Statement (EIS: V-77, 78)—been used instead of the May 1981 figure. Use of that date is stipulated by P.A. 450's definition of "initial assessed value" as the most recent assessed value of property in a development area at the time the ordinance establishing a tax increment financing plan is adopted; because assessed values in Michigan are officially set when rolls are equalized on the first Monday in May, the adoption of the tax increment financing ordinance in April 1982 resulted in the use of May 1981 as the date for the initial SEV for the project area. This plan illustrates only too well the primary disadvantage of tax increment financing; that the municipal general fund and other taxing districts receive little of the benefits from the new investment.

EVALUATING DEVELOPMENT STRATEGIES

Without state-provided economic development programs and federal funding, it is improbable that Detroit would have embarked upon or been able to complete the General Motors project in accordance with General Motors' specifications and deadlines. In this sense, then, the programs can be considered effective in accomplishing their goal. However, the more difficult question relates to the impact of the programs on the corporation's plant location decisions: Did they attract investment where it would not otherwise have gone, and were *all* of them necessary to attract that investment? The existing literature is solidly pessimistic: Most corporate locational decisions seem to be made without reference to public sector incentives.

One reason for the limited impact of public inventives on plant location decisions is that they affect a relatively small proportion of the total costs of building and operating a typical new industrial facility. Much more significant than state and local taxes in a firm's calculation are construction costs, land prices, and labor costs (Hammer, 1974). The Council for Urban Economic Development (1981: 37) reported that many firms have labor costs that are 20 times as large as state and local tax payments, observing that a 2 percent wage differential can have as much impact on a company's "bottom line" as a 40 percent differential in taxes. Labor costs have also frequently been cited by surveys of businessmen as a major factor in relocation decisions (Mandell, 1975).

Another factor reported by survey respondents is the inadequate space for expansion available in older cities. Schmenner's (1980) study of survey responses from plant officials in New England and Cincinnati indicated that the "overwhelming reason why plants relocate is to secure additional space" (1980: 455). Such moves often are necessary to utilize new technology in plant or equipment design, or to minimize energy costs in plant operations. Schmenner (1980) concluded that land assembly and site preparation is the most important economic development policy that cities can pursue, and urged wider use of eminent domain powers to accomplish this end. However, Reagan's budget cuts have severely curtailed the programs that Detroit used in site acquisition and preparation for the Poletown project.

The empirical work relating state and local development and fiscal policies to industrial location decisions is also discouraging. Oakland (1978) reviewed three regression studies assessing the effect of tax rate differentials on the distribution of industry within a metropolitan area, and found no evidence that "intra-urban location decisions of business firms are significantly affected by fiscal considerations" (Oakland, 1978: 28) Pascarella and Raymond (1982) report a similar conclusion for industrial revenue bonds in Ohio. In general, then, evidence from surveys of businessmen and aggregate regression analyses suggests that public incentives have little impact on private investment decisions *generally,* although they can make a difference in a specific instance.

THE POLITICS OF THE CORPORATE SURPLUS

Our research on the politics, financing, and implementation of a project to build a new General Motors assembly plant in Detroit provides evidence of the dominant position of the private sector in economic development policies, a position stemming from the public competition for private investment. It portrays the essential role played by federal and state governments in enabling and financing urban industrial renewal, and indicates the consequences of the lack of policy review by those governments.

We do not claim that this case study is typical. General Motors is a corporate giant whose multinational scope of operations maximizes its capital mobility; Detroit's economic situation is worse than that of many northern industrial cities, making its officials especially desperate to retain jobs and increase tax revenues. Moreover, the access of Detroit's mayor, Coleman Young, to high-level Carter administration officials during the formative stages of the project facilitated extensive use of federal funds. Rather, the case highlights how severe fiscal problems, maximum utilization of intergovernmental funding and enabling mechanisms, and the capital mobility of a large corporation contribute to an economic development policy that facilitates the extraction of the maximum corporate surplus.

GM DEMANDS

The Detroit GM project represents the city's response to an offer in April, 1980 from the head of GM's real estate division to construct a new plant in the city to replace two obsolete facilities scheduled to close in 1983, provided that city officials could find an adequate site in time for the new plant to begin production in the summer of 1983. Construction of the new plant would be part of General Motors' $40 billion national capital investment program, involving the replacement of older plants with new ones capable of producing front-wheel drive vehicles, which were expected to be 80 percent of the corporation's production by 1984. Throughout the project, conditions articulated by GM were a determining factor in policy formation—conditions that city officials could either accept or risk the corporation taking its money and its jobs elsewhere. Detroit's fiscal and economic situation created pressures that reinforced those stemming from GM's mobile capital in leading city officials to not only accept the corporation's conditions, but to stretch the boundaries of a number of financial programs and state enabling legislation to put together a project plan that met the corporation's technical requirements, minimized its share of project costs, and complied with its demanding time schedule.

The corporation's influence was first evident in the site review process, during which a committee of city officials and GM representatives evaluated alternative plant sites. All but one of nine alternative locations failed to meet GM's site criteria, incorporated in a standard plant design utilized for all of the corporation's new facilities, most of which were in semirural locations rather then densely populated urban areas. The site chosen, an area of approximately 450 acres on the Detroit-Hamtramck border, included a large abandoned Chrysler plant (Dodge Main) as well as an older Detroit neighborhood, known as Poletown, of some 3,500 residents. The presence of the Dodge Main factory meant that a large proportion of the land could be made available without relocating residents; construction of a new plant would also solve the problem of finding a use for the massive abandoned Dodge Main complex. Boundaries of the project area, labeled the Central Industrial Park, were approved by the Detroit City Council in July 1980, and preparation of the project plan was initiated by the staff of the Community and Economic Development Department.

The contributions of the city staff were limited by GM's technical requirements for plant design and physical layout, as they had been in the site review stage of the project planning. GM's prototype plan specified the assembly plant's size and shape, the configuration of rail yards and docks (which in turn dictated the location of other facilities on the site), and the inclusion of a power plant, pump house, waste water treatment facility, storage space for components and completed automobiles, and parking space for 3,000 workers in each of two shifts.

The city's concern with speed in the preparation and implementation of the plans for the project can be attributed to the rigid timetable for the project set by General Motors, which was dictated by the corporation's objective of completing construction of the plant by May 1983. The project plan adopted in October 1980 outlined five phases of property acquisition and clearance, with the transfer of the first cleared section to General Motors scheduled for May 1981 and the last for June 1982. GM also demanded the "maximum allowable tax abatement" under P. A. 198, the vesting of title to the entire state in the city of Detroit by May 1, 1981, offered a purchase price of $18,000 an acre, and set forth detailed specifications for site preparation and the provision of roads, street lighting, and other utilities.

CITY RESPONSE

The General Motors deadlines and the implied threat of building the plant elsewhere if the conditions were not met were constantly mentioned in city council and community meetings on the project. Special task forces were established to arrange financing and to ensure the timely implementation of components of the project plan by the numerous local government agencies whose cooperation was necessary; the implementation task force established the project as a top policy priority and representatives from affected agencies saw to it that time-consuming standard operating procedures were averted. City officials, particularly Mayor Coleman A. Young and development director Emmett S. Moten, Jr., were in constant contact with HUD officials to ensure that the various components of the financial package were in place for city council approval of the entire plan in late October 1980. The city's major contributions were the development of financing plan for the project, incorporating estimated costs and projected funding sources, the formulation of plans for relocation of residents and businesses in the project area, and site clearance.

The financial plan, as noted earlier, brought together a variety of intergovernmental grants and loans, primarily from the federal Department of Housing and Urban Development. HUD policies and regulations were also significant in shaping relocation plans, which were based on the Federal Uniform Relocation Assistance and Real Property Acquisition Policies Act of 1970 (referred to heareafter as the Uniform Act). In order to expedite clearance of the area as well as respond to concerns expressed by area residents and the citizens district council, additional benefits and special procedures were established for the General Motors project. These additions included (1) a $1,000 bonus for residents and businesses moving within 90 days of receiving a notice of displacement; (2) compensation to homeowners for higher interest rates on new mortgages; and (3) compensation to homeowners whose property taxes increased as a result of relocation. These additions increased relocation costs by more than $2 million. Relocation staff also attempted to be "generous" in acquisition awards and in defining "housing of last resort," in which HUD permits replacement housing payments in excess of the $15,000 maximum set by the Uniform Act.

The General Motors schedule prevented extensive study by the council of the project plan or of citizen concerns voiced at public hearings and council meetings. Although the staff of the city planning commission presented the council with recommendations throughout its deliberations on various aspects of the project, these were usually presented to council members on the day a vote was required, and were thus not incorporated in policies (the deadlines and threat of relocation were often cited by city officials as not allowing any postponement of council action). Each council approval of applications for grants and loans, of the project plan, of tax abatement, and so on contributed to project momentum and was used to justify subsequent actions; for example, during council discussion of the final component of the project, establishing a tax increment financing authority, CEDD officials reminded members that this action was required by the Interlocal Agreement with Hamtramck, which the council had approved the previous year.

Once the project plan had been approved by the council, staff in the property acquisition and relocation divisions of CEDD moved quickly to send out formal offers to property owners and provide relocation assistance. In 17 months, 3,500 individuals and 162 businesses were relocated (some businesses did not reestablish elsewhere) and acquisition and demolition of 1,500 structures was completed. Extra staff assigned to the relocation division, the availability of federal funds for the relocation incentives described above, and the provisions of the "quick take law" expedited the acquisition and relocation process. Demolition was facilitated, according to city officials, by assigning an "umbrella contract" to one contractor, which served as the city's site manager and subcontracted specific aspects of demolition and site preparation; this, they claimed, provided greater flexibility and eliminated such potentially delay-inducing bureaucratic procedures as obtaining city council approval for each contract.

POLICY REVIEW IN THE COURTS

The rapid pace of project planning and the multiple sources of financing and authority meant that no review of such issues as the total cost of the project and the taking of private property for the use of other private interests occurred. Indeed, the only state-level review of the development plan occurred during legal challenges by opponents of the project, and then only after implementation of the project plan had begun. The Poletown Neighborhood Council, an organization formed in July, 1980 to represent those residents of the project area who did not want to relocate, filed a suit in Wayne County Circuit Court in November, 1980 charging that city officials had abused their authority by giving inadequate consideration to alternative sites or plants, and that their actions did not serve a "public purpose" as required by state law, because a private corporation would be the primary beneficiary of the project. The legal bases for the challenge were grounded in the state constitution's definition of eminent domain and the authority granted to the city and its Economic Development Corporation by P.A. 338 (the enabling legislation) and P.A. 87 (the "quick take" law). The city received a favorable ruling from Wayne County Circuit Judge George T. Martin.

The Poletown Neighborhood Council appealed Judge Martin's decision to the Michigan Supreme Court, and obtained a temporary injunction from the court in February, 1981 (Poletown v. Detroit, 1981). The injunction put pressure on the court to render a prompt decision, as city lawyers emphasized the city's need to have title to all property by the May 1 GM deadline.

Less than two weeks after hearing oral arguments, the state Supreme Court rendered a 5-2 decision rejecting the Poletown appeal. In rejecting the charge that the project served primarily the private interest of General Motors, the court majority noted that the power of eminent domain "is not to be exercised without substantial proof that the public is primarily to be benefited"; but the five justices believed the city had presented such proof. The two dissenting justices, however, expressed concern that the court's ruling could set a precedent for "the most outrageous confiscation of private property for the benefit of other private interest without redress."

POLICY DISCRETION AND PRIVATE INTEREST

Less than two years transpired between the public announcement of the project and the construction of the assembly plant. During that time, $200 million in public (mainly federal) funds was spent on site preparation, and the application and interpretation of four state enabling laws (P.A. 338; P.A. 450; P.A. 198; P.A. 87) significantly extended the eminent domain powers of local government and granted discretionary taxing authority without concern for broader policy implications.

Both state and federal legislation allowed Detroit officials maximum discretion in combining multiple financing sources and enabling authority, and in judging the value and feasibility of the project. City officials, however, were the least capable of rendering an objective and comprehensive evaluation for two reasons: (a) Detroit's economic situation created strong pressures to maximize employment and tax revenues in any way possible; and (b) as the smallest geographical unit of government, the city was at the greatest disadvantage in relation to the private sector power stemming from the mobility of corporate capital.

The ultimate consequence of the confluence of extensive intergovernmental funding sources and grants of economic development authority, favorable political ties with federal agencies, and a desperate need for jobs and tax revenues in Detroit was a situation that maximized private sector influence in the policymaking process. This influence resulted in the extraction of a large corporate surplus. City officials had no way of limiting the public-provided incentives flowing to the corporation. Indeed, success was defined for them as achieving

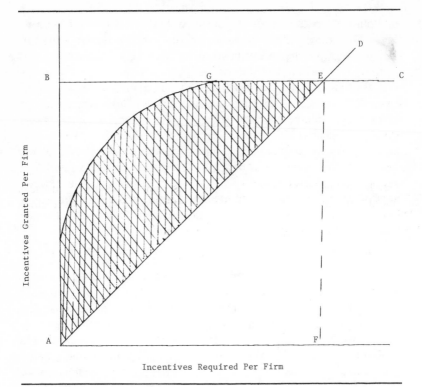

Figure 13.1 The Corporate Surplus

the maximum incentive for the corporation from whatever source was available. Grants of policy discretion to the city had the effect of maximizing the ability of General Motors to extract resources from the public sector. A territorially small government can easily become a "company town," acting as effective advocate for the corporation.

THE CORPORATE SURPLUS

In the marketplace, a *consumers' surplus* exists because there are consumers who are willing to pay more than a producer's asking price (Stokey and Zeckhauser, 1978: 150). Competition in the marketplace reduces the asking price and, in the limit, maximizes the size of consumer surplus.

In an analogous manner, competition among governments in the public economy for private capital investment results in a corporate surplus. While individual negotiations between government and corporations have the potential for exactly matching the needs of corporation with the appropriate public incentives, this rarely happens.

The corporate surplus may be better understood by examining Figure 13.1, where the incentives granted by a municipality are graphed against the incentives required by firms. This relation is labeled "AGC." The upper limit of incentives is set by the state's enabling statutes labeled "BC." The ideal match that could be achieved in a world of perfect information is labeled "AD."

Corporations planning relocation have motive to conceal the incentives required to influence their decisions; however, where the incentive requirements are relatively low, city officials should be able in many cases to ascertain this fact. Automobile suppliers, for example, have lower incentive requirements than some other business operations since they are tied so completely to the location decisions of the automobile companies. Hence the city will not invariably grant maximum incentives. However, as the incentives necessary grow, the incentives granted quickly reach the statutory limit as an asymptote. The form of this curve stems from the asymmetry in the knowledge relationship between firms and municipalities.

The corporate surplus is the area between the incentives granted, curve AC, and the ideal match, AD, constrained, of course, by the statutory limits. The total value of the incentives provided is the area of the figure AGEF. The value of the incentives necessary to influence corporate decision making is represented by the area of the triangle AEF. The corporate surplus is the value of total incentives less the value of necessary incentives, AGEF to AEF. This is shaded on the diagram.

CONCLUSIONS

In this chapter, we have used the concept of the corporate surplus to analyze city economic development policies. We have described in some detail the specific economic development policies used by the city of Detroit in negotiating the location of a new automobile assembly plant with General Motors, and have indicated how city officials made and implemented specific project decisions.

We could emphasize several conclusions from this case study—for example, the ability of city officials to impose a policy settlement in the face of rancorous opposition, the entrepreneurial skills necessary to complete the project, and the close collaboration of Detroit officials with the national Democratic administration. We will stress, however, three important conclusions in relation to the generation of the corporate surplus.

(1) In the area of economic renewal, private corporations possess important power resources as a consequence of the mobility of capi-

tal. The city's major interest in promoting industrial renewal stems from fiscal strain. If service needs were not so rapidly outrunning the ability of the urban tax base to produce revenue, cities would not be so interested in "chasing smokestacks." The power of corporations to extract concessions from city governments is rooted in a simple and above-based exchange relationship: The corporations have something the city wants, and the corporation is not required to make the desired object available to the city. Hence the city must pay to get it. It must produce appropriate policies to the limit of its legal authority. Of course, the incentives that the city can provide a corporation that is in the process of making a locational decision are not likely to play much of a role in the decision itself, because such public sector incentives have been shown in study after study to comprise but a minor ingredient in the mix of factors influencing the locational decisions. This, however, is beside the point, simply because several sites are likely to fulfill the corporation's criteria. If it has more than one potential site, the corporation holds all the chips and will then be able to extract up to the legal limit.

Corporate power, at least in the arena of urban industrial renewal, stems solely from the mobility of capital and the relatively small size of urban jurisdictions. Because American municipalities possess significant policymaking authority, they are forced to compete with one another in order to attract the commercial and industrial concerns that their leaders see as critical in building and maintaining a healthy private sector base for public activities. Without local discretion, there would be no "corporate blackmail." Yet without local discretion, there would be no local self-government.

It is, of course, possible that private elites have built and maintained such an arrangement for political advantage (Stone, 1981). We cannot accept this thesis. The tradition of local autonomy predates the modern corporate structure by centuries; local self-determination is deeply rooted in the American culture. The modern managers of multinational corporations did not establish this institution any more than they established the nation-state, an international arrangement that redounds to their advantage today. Yet only the most inept businessman would fail to exploit this system to his advantage.

(2) Discretionary policies are not always discretionary. State governments have in recent years enacted all sorts of policies that enable municipalities to grant incentives to promote economic development. We have reviewed a number of these programs. In almost every case, the incentive programs are intended by state government to be discretionary; no municipality is required to grant tax abatement

or issue industrial revenue bonds. In almost no case is any program truly discretionary. By virtue of the asymmetrical power relationship that characterizes the relationship between corporations contemplating relocation and municipal governments, local governments lose the ability to choose. They cannot match the level of public incentives to the minimum necessary to affect the corporate locational decisions because municipal officials do not know what those requirements are. On the other hand, every corporation knows exactly how far a municipality is prepared to go: to the statutory limits of the enabling legislation. So long as the corporation is indifferent between two or more sites in two or more jurisdictions, or can convince municipal officials that it is, it will normally be able to extract maximum incentives.

The amount of locational incentives provided by a municipality beyond the minimum necessary to retain or attract a facility we have termed the *corporate surplus*. The existence of the corporate surplus stems from the knowledge asymmetry that exists between municipal officials and corporate executives. By failing to establish adequate standards for use of the various incentives that have been made available, state governments have put municipalities in the position of having little in the way of resources to oppose the demands of corporations for maximum deployment of statutory incentives. In fairness, it must be emphasized that oftentimes municipal officials have been the most consistent advocates of the economic development enabling legislation provided by the states. Municipal governments have bestowed on themselves discretionary policies that make it almost impossible for them to exercise any discretion.

(3) Policy review of local government actions rarely occurs. In the matter of the Detroit Central Industrial Park, neither the federal government, which facilitated the project through a complex arrangement of loans and grants, nor the state government, which readily passed the necessary enabling legislation, engaged in anything approaching a comprehensive evaluation of the project. The city of Detroit did an evaluation, but, as we have noted above, the municipality had lost its policy prerogatives because of the structure of the situation. There are two aspects of the granting of locational incentives that ought to be considered by higher levels of government. These are, first, the question of the total social benefits of the incentives and, second, the control of the corporate surplus.

The city of Detroit spent some $200 million dollars for acquisition, demolition, relocation, and site preparation for the Central Industrial Park. Had the plant been built in an open field that was relatively accessible to transportation, the public could have saved the $200 million (plus interest). Note that Detroit had to spend this money "up front," prior to negotiating any tax abatements or financial arrangements for the construction of the plant itself. This site clearance was necessary just to get in the ball game, to be able to compete on a more or less equal footing with other communities.

Only at higher levels of government can the issue of the corporate surplus be dealt with. The issue of a corporate surplus arises because of the mobility of capital, the power of communities to grant locational incentives, and the differences in access to information that government and corporations possess. If the unit of government is large enough to encompass all potential sites for a facility, the issue of the corporate surplus is moot. It an increasingly interdependent world, it is highly possible that even the nation-state cannot deal with the problem of corporate surpluses; certainly municipalities cannot.

REFERENCES

BLUESTONE, B. (1983) "Economic turbulence: capital mobility vs. absorptive capacity in the U.S. economy." Presented at the Meetings of the American Association for the Advancement of Science, Detroit, Michigan, May.

CLARKE, S. (1982) "The private use of the public interest." Presented at the Southwestern Political Science Association, San Antonio, Texas, March.

COOK, T. (1982) "The courtship of capital." Presented at the meeting of the American Political Science Association, Denver, Colorado, September.

Council for Urban Economic Development (1978) Coordinated Urban Economic Development: A Case Study Analysis. Washington, DC: Author.

———(1981) State Actions to Stimulate Development: A Catalog. Washington, DC: Author.

FRIEDLAND, R. (1982) Power and Crisis in the City. London: Macmillan.

HAMMER, A. M. (1974) The Industrial Exodus from the Central City. Lexington, MA: D.C. Heath.

KATZNELSON, I. (1976) "The crisis of the capitalist state: urban politics and social control," in W. Hawley et. al. (eds.) Theoretical Perspectives on Urban Politics. Englewood Cliffs, NJ: Prentice-Hall.

KIRBY, A. (1982) The Politics of Location. London: Methuen.

LONG, L. and D. DeARE (1983) "Metropolitan-nonmetropolitan industrial changes and population redistribution." Presented at the American Association for the Advancement of Science, Detroit, Michigan, May.

LUND, L. (1979) Factors in Corporate Location Decisions. New York: The Conference Board.

MANDELL, L. (1975) Industry Location Decisions: Detroit Compared with Atlanta and Chicago. New York: Praeger.

MAURER, R.C. and J.A. CHRISTENSON (1982) "Growth and non-growth orientations of urban, suburban, and rural mayors: reflections on the city as a growth machine." Social Science Quarterly 63 (June): 350-358.

MOLOTCH, H. (1976) "The city as a growth machine: toward a political economy of place." American Journal of Sociology 82 (September): 309-332.

OAKLAND, W. H. (1978) "Local taxes and intra-urban industrial location: a survey," in G. Break (ed.) Metropolitan Financing and Growth Management Strategies. Madison: University of Wisconsin Press.

Poletown Neighborhood Council v. City of Detroit (1981) 410 Mich. 616; 304 N.W. 2d 455.

PEIRCE, N. R. and C. STEINBACH (1980) "Reindustrialization—a foreign word to hard-pressed American workers." National Journal October 25: 1794-1789.

RUBIN, I., C. H. LEVINE et al. (1981) "States' role in local fiscal stress: observations for six localities." Prepared for the Midwest Political Science Association Annual Meeting, Cincinnati, Ohio, April.

SCHMENNER, R. W. (1980) "Industrial location and urban public management," in A. P. Solomon (ed.) The Prospective City. Cambridge MA: MIT Press.

STOKEY, E. and R. ZECKHAUSER (1978) A Primer for Policy Analysis. New York: Norton.

STONE, C. (1981) "Social stratification, non-decision-making, and urban policy." Presented at the Southwest Political Science Association Meeting, Dallas, Texas, March.

VAUGHAN, R.J. (1979) State Taxation and Economic Development. Washington, DC: Council of State Planning Agencies.

14

A Tale of Two Cities:
A Case Study of
Urban Competition
for Jobs

JOHN P. BLAIR and BARTON WECHSLER

Chicago—September 27, 1982: "International Harvester today announced that its U.S. truck assembly operations will be consolidated at its Springfield, Ohio plant. The decision means that the company's Fort Wayne, Indiana truck assembly plant will be phased out. The transfer of heavy truck production to Springfield is expected to begin in November."

The economic problems of International Harvester and the repercussions felt in the communities of Fort Wayne, Indiana and Springfield, Ohio provided an example of the postindustrial shift away from manufacturing and the interurban competition for jobs that has arisen from this shift (see Bluestone and Harrison, 1982, for a broad discussion of this problem).

SETTING THE STAGE

By 1981, in the face of continuing losses, International Harvester (IH) had accumulated a debt of approximately $4 billion and was unable to continue operation without a major restructuring of its internal operations and debt obligations. The company's alternatives were to restructure its loans (favored by the Chairman and Chief Executive Archie R. McCardell) or to file for protection under the bankruptcy laws (favored by IH President Warren Hayford). Just before Christmas 1981, after five hours of debate, the board of directors voted to support McCardell's loan restructuring strategy.

After internal disagreements culminating in the resignations of both McCardell and Hayford, a $3.5 billion restructuring proposal was approved by the board. International Harvester also provided for conversion of up to $350 million in loans and interest into equity through the purchase of stock by independent IH dealers and concessions from suppliers totaling about $50 million (Marsh and Saville, 1982: 38-42). In addition, the company planned to consolidate its operations by selling its construction division and numerous small operations including the axle assembly plant in Fort Wayne and the remanufacturing facility in Atlanta. Plants throughout the United States and in Europe were discreetly offered for sale.

A TALE OF TWO CITIES

The decision to consolidate U.S. truck operations grew out of the overall restructuring program introduced under the new leadership of Chairman Louis W. Menk and President Donald D. Lemmoy. The initial proposal came from an executive at the Fort Wayne plant who suggested that consolidation might be helpful in securing the loans needed for the restructuring plan. IH executives were receptive to the idea in principle, but made no decisions at that time.

Rumors about a potential plant closing came to the attention of the Springfield Area Chamber of Commerce's Industrial Development Committee through information received by Springfield's Representative Clarence Brown. In response to this potential threat, a group of local officials made a trip to IH headquarters in Chicago in order to discuss the situation. At that time, company officials expressed uncertainty about their plans for truck operations. Nevertheless, the Springfield delegation was concerned about the continued operation of the IH plant in Springfield and suggested the possibility of incentives to keep the plant open. Specifically, the Springfield contingent suggested the possibility of an arrangement in which Springfield would buy the assembly plant and then lease it back to IH. This proposal would provide liquidity for the debt-ridden company and was, therefore, attractive to IH. However, corporate officers expressed a reluctance to finalize a purchase-lease back agreement since definitive plans for truck operations and for maintenance of the Springfield plant had not been made.

By March 1982, International Harvester was seeking financial assistance from both Fort Wayne and Springfield and from their respective state governments. IH representatives claimed that there were cost reductions that could permit continued operation in both facilities. While officials representing Fort Wayne and Springfield were developing separate funding packages, there was substantial contact and communication between them. Frequent meetings and consultations resulted in both communities putting forward the idea of a purchase/lease-back arrangement that would reduce IH's cost of operation and provide short-term liquidity.

During the spring of 1982, both Springfield and Fort Wayne worked on the technical details of their proposals. Despite assurances that the two cities were not in competition and that the continued operation of both plants was still possible, newspaper reports were now presenting the choices as mutually exclusive and stressing the competitive position of the two communities. There is no evidence that IH officials encouraged an adversary relationship between the cities, but that perception did emerge.

In July, IH officially announced that it would be unable to continue operations at its existing locations and that one plant would have to be closed down. Assurances were again offered that there was no competition between the two cities and at no time did IH state that it would continue operations in the city making the best subsidy offer. Nevertheless, both cities worked to develop the most attractive packages possible in time for IH's September decision.

SPRINGFIELD'S OFFER

Springfield officials structured their proposed purchase-lease back around a newly created Community Improvement Corporation (CIC). The structure of this arrangement is shown in Figure 14.1. Under the terms of the proposal, CIC would pay the appraised value for the IH truck plant and then lease the facility back to the company. The purchase money would be lent to CIC from private and state sources. The rent paid by IH for the use of the plant would then be used by CIC to repay the loans made to it. The basic structure of this proposal remained relatively constant throughout the negotiation period. However, securing the necessary financial commitments from lenders proved difficult.

When the purchase-lease back was first proposed, it was assumed that financial backing would come entirely from private sector sources. Given IH's deteriorating financial status, private interest in the proposition was limited. Springfield and state officials then attempted to secure the participation of the State Teachers Retirement Fund (STRF), but again to no avail. Teachers and officials of the STRF feared that possible default would place the loan in jeopardy. Ultimately, private financial institutions became the primary source of funds for the CIC. Through the leadership of Huntington National Bank, a coalition of eleven banks and savings and loans agreed to lend approximately $18.4 million to the CIC. The participation of these institutions was largely motivated by their fear of the potential repercussions associated with closing of the IH plant in Springfield. Not only would a plant closing reduce deposits, but more importantly, it would jeopardize the security of real estate mortgages and lower

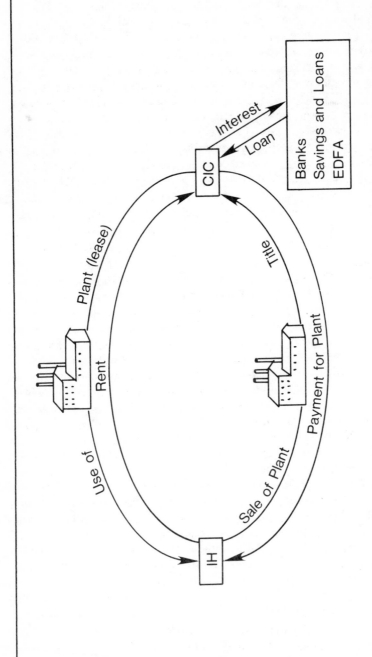

Figure 14.1 The Springfield Sales/Lease-Back Arrangement

housing prices. In addition to the effects on lenders, it was estimated that for each job lost at the Springfield plant, an additional 1.4 jobs would be lost elsewhere in the community.

While Springfield officials were discussing loan possibilities with the financial community, the Ohio Department of Economic and Community Development was beginning implementation of the Economic Development Finance Act (EDFA), a new loan program designed to assist financially troubled Ohio companies. The principal EDFA criteria were (1) that EDFA assistance was required to create or preserve jobs in Ohio; (2) that the proposed project would fail without public participation; and (3) that a maximum of private participation had been secured. EDFA was intended to serve as gap financing and would provide highly leveraged public monies. After a period of negotiation, EDFA administrators agreed that a loan of $9.2 million to the CIC would be appropriate. Thus, of the $27.6 million plant purchase price, $18.4 million was to be secured from private financial institutions and $9.2 million from the state of Ohio.

An additional problem emerged when the prospective private lenders expressed new concerns about the riskiness of the venture. Although the state of Ohio had agreed to guarantee up to 50% of the loans against default, some of the lenders now wanted larger guarantees. Without these guarantees, the lenders claimed it would not be possible for them to make the necessary loans. In the end, the state agreed to guarantee 85% of the value of the loans. Thus, although private sources provided nearly two-thirds of the funds for the purchase of the Springfield plant, the state of Ohio assumed almost the entire risk of an IH default on the lease.

The rent IH was to pay to the CIC for use of the plant was based on the cost of the two loans. On the EDFA loan, IH would pay 2% for the first ten years and then 10% for the second ten years. On the bank loans, IH would pay the prime rate plus 1% for the entire twenty-year period. The total cost to the company was well below what IH would have had to pay in the private market for a comparable facility. Yet, the lending institutions were receiving market or near market rates of return on their loans in light of the safety provided by the state loan guarantees. Moreover, the lease payments from IH would be sufficient to enable CIC to pay off its debts to the lenders.

FORT WAYNE'S OFFER

Fort Wayne did not begin to assemble its financial package until almost six months after Springfield had made its initial purchase/

lease-back proposal to International Harvester. Unlike Springfield's subsidy plan, which remained unchanged in value, the Fort Wayne package grew from $9 million to $18 million and then to $31 million.

Fort Wayne's first proposal was a $9.2 million loan guarantee. Later, the state of Indiana committed $4 million and Allen County agreed to provide $2 million. Fort Wayne offered $3 million more. These funds would be used to arrange a purchase-lease back of the Fort Wayne plant similar to Springfield's. Under the plan, IH would sell its plant to a private investor who would lease the facility back to the company through a local government agency. The funds used for the purchase-lease back would be secured by the IH parts center and by IH's pledge of unobligated machinery and equipment.

Fort Wayne kept adding to its financial package in an attempt to at least match Springfield's proposal. Fort Wayne's package included an extra $8 million in low interest loans from public and private sources. These loans would be used to upgrade various plant operations and facilities, making the Fort Wayne operation more efficient.

Fort Wayne continued to augment its proposal until they had amassed approximately $31 million in loans and lease guarantees including funds from local banks and federal block grants. A last-minute newspaper campaign even induced one-fifth of the work force at the Fort Wayne plant to agree to defer 15 percent of their wages if the plant were kept open. (However, some IH officials believed such an arrangement would violate union contracts.)

INTERNATIONAL HARVESTER'S DECISION

The decision reached by IH was based on both short- and long-run considerations. In terms of the start-up costs associated with closing one plant and consolidating operations at the other, Fort Wayne appeared to have an advantage. In order for the Springfield plant to assemble the heavy trucks being produced at Fort Wayne, extensive and costly retooling would be necessary. On the other hand, transferring operations from Springfield to Fort Wayne would be much less costly. On a short-run basis, it appeared that a reorganization that continued operations in Fort Wayne would be most efficient. However, long-run considerations favored the more modern Springfield plant that, once retooled, could operate at a lower average cost than the Fort Wayne facility.

Ultimately, IH announced its decision to close the Fort Wayne facility and consolidate its truck assembly operations at Springfield. The company's press release stated that the financial incentives offered by the two cities were so comparable as not to be a factor in the final decision. However, an IH spokesman indicated that the certainty and extent of commitment of the Springfield package compared to the Fort Wayne offer was a consideration in the final decision. According to IH officials, it was primarily the comparative efficiency of the Springfield operation that was the deciding factor. Other benefits associated with consolidation at the Springfield location included an automated high-rise warehouse for inventory storage, a more efficient materials delivery system, and a better plant layout.

Springfield residents greeted the decision by IH with a collective sigh of relief. Not only would the current jobs at the Springfield plant be saved, but over time IH planned to increase employment as operations were shifted from Fort Wayne. The feelings of relief were, nonetheless, mixed with concern about the future of International Harvester. A feeling of crisis not resolved, but at least postponed, was widespread. Some observers noted that even if IH later became bankrupt, at least the CIC now owned the plant so it might continue operations with limited disruptions.

About a year after the decision, Springfield still has not regained economic vitality. The unemployment rate is high, as is the housing vacancy rate. However, the outcome of the plant decision coupled with publicity from a *Newsweek* cover story devoted to Springfield and plans for a made-for-TV movie based on the *Newsweek* article have generated a sense of optimism and community. On the other hand, recent demographic shifts have moved the city of Springfield into the Dayton SMSA, and Springfield residents are concerned that this shift could weaken the emerging sense of identity.

Fort Wayne reacted to the announcement with resolution to revive local economic development efforts. Their strategic planning was predicated on the belief that Harvestor's pull out was only one of four problems facing the region's economy. Gathrie (1982) described three other reasons for Fort Wayne's decline:

(1) A poor industrial base. About one-third of the area's employment was in manufacturing.
(2) General geographic decline. The north-central states had the poorest economic performance during the 5 years.
(3) The agribusiness slump. The dramatic loss in farm income reduced the metropolitan area's hinterland market as well as the export base.

Several initiatives were undertaken in order to attract industry. Steps included reappraisal of existing programs, reorganization of existing economic development groups into the Greater Fort Wayne Chamber of Commerce, and selection of the Fantus Corporation to assist in marketing Fort Wayne to potential new employers (Fantus, 1982). In addition, state and local officials reviewed the program of economic development incentives (tax abatements, low interest loans, etc.) available to the community in its efforts to retain jobs.

A THEORETICAL PERSPECTIVE

Problems of community reaction to plant closings and threats of plant closings have sparked a burgeoning literature. Dotson (1983: 359) summarized some of the common findings from a variety of treatises:

(1) Community economic trauma is occurring in significant numbers.
(2) Businesses are reluctant to reveal plans or discuss particulars about decisions.
(3) Causes of shut-downs are multiple including industry-specific factors, and macroeconomic fluctuations.
(4) Ripple effects of shut-downs are significant.
(5) Good data is sparce.
(6) Damages persist for extended periods although the immediate impact is mitigated by state and local assistance programs.
(7) Shut-downs are an appropriate area for policy intervention.

The Springfield-Fort Wayne case supports the first 5 generalizations. Not enough time has passed to fully report the long-term impacts in Fort Wayne. The seventh generalization is not strictly supportable because it involves policy/value judgments, but many officials and citizens in Fort Wayne would agree with it.

One of the unique wrinkles in this case was the "head to head" interurban competition. How can we explain the competition between Springfield and Fort Wayne to maintain the IH plants? An appealing behavioral hypothesis is that officials representing both cities were attempting to purchase well-paying jobs and related spillover benefits from jobs for their residents. Although the complexity of the financial packages and negotiations may obscure this essence, the two cities were competing in a *market for jobs* (Blair et al., forthcoming). The market was characterized by a monopolistic seller (International Harvester) and two buyers (Springfield and Fort Wayne).

Table 14.1 illustrates this market in a game theoretical framework. In this illustration, we will assume that residents of both cities receive positive net benefits from job creation. However, residents of City B receive greater benefits from a location at B than residents of A would receive from a location at A. The payoff to the firm from locating in either A or B is the same, $25. The payoff matrix represents benefits before sidepayments have been made. In this case, B is the socially optimum location because total social benefits are greater at B. But what incentives are there to induce the firm to select the socially optimum location? In the market for jobs, residents of B could outbid residents of A, thus providing the appropriate system of incentives.

While the market for jobs can improve locational decisions and create local jobs, it may also contribute to national job creation. Four possible outcomes of a well-functioning market for jobs are summarized in Figure 14.2. (A well-functioning market can be said to exist if the communities bid (1) in proportion to the benefits they receive from a location and (2) enough to make a firm's operation profitable if there are sufficient local social benefits from a location.) The payoffs shown in the matrix can be redistributed through local competition in the market for jobs to generate the outcomes.

A illustrates a situation in which a well-functioning market for jobs will cause net new jobs to be created in the nation, not just a zero sum game. This can be seen because in the absence of urban development subsidies, the firm could not operate profitably anywhere. The market for jobs will also induce the firm to locate in the city where the external benefits from the creation of jobs is greatest, 2 in this illustration.

B and C represent intermediate cases. In case B, neither location is superior to the other so a well-functioning market for jobs will not improve locational choice because the two sites have equal external benefits. However, if the cities transfer to the firm a portion of their external benefits, there will be a national (as well as local) job creation impact. In case C, the firm would operate at 1 in the absence of economic development sidepayments. However, because of the greater importance of job creation in City 2 (perhaps 1 has no un-

TABLE 14.1 The Market for Jobs

Recipient	Location	
	A	B
Firm	$25	$25
Residents A	10	0
Residents B	0	25
Net social benefits	$35	$50

LOCATIONAL IMPACT

NATIONAL EMPLOYMENT IMPACT

Job Creation

A   Efficiency

	Site	
	A	B
Firm	–5	–5
City A	10	0
City B	0	25
Net Social Benefits	5	20

B   No Locational Effect

	Site	
	A	B
Firm	–5	–5
City A	10	0
City B	0	10
Net Social Benefits	5	5

No Net Job Creation

C

	Site	
	A	B
Firm	10	5
City A	10	0
City B	0	25
Net Social Benefits	20	30

D

	Site	
	A	B
Firm	10	20
City A	10	0
City B	0	10
Net Social Benefits	20	30

Figure 14.2   Possible Outcomes of a Well-Functioning Market for Jobs

employment and 2 has high unemployment), the net benefits of location will result through the operation of the market for jobs.

D shows a case in which the market for jobs may be purely distributive rather than contributing to either an efficient location or job creation. The creation benefits to both 1 and 2 are equal and the firm could operate profitably in the absence of any local subsidy. Assume each community bids in proportion to the social benefits residents would receive if they were selected as the firm's location. The firm would make the same choice that it would have made in the absence of a market for jobs, City 2. There will be an effect on national resource allocations but wealth will be transferred from residents of the successful bidder city (City 2) to the firm. This model appears to represent the IH case. Both Springfield and Fort Wayne would have received approximately equal benefits from a decision to maintain their local plant. However, IH received greater benefits from selecting Springfield. In either location, operations would have been profitable.

Although a market for jobs can be an effective approach in job creation and site creation, it is currently inefficient. One reason for the inefficiency is that the market for jobs is, at present, an incipient market characterized by implicit and ill-defined agreements. The poorly organized and heterogeneous nature of the market is one reason that these local job creation efforts have not heretofore been recognized and analyzed as a market. A recent HUD study found that the value of employment subsidies ranged from $0 to $100,000 in some EDA projects. The large spread indicated an inefficient market.

There are several important reasons why the market for jobs is inefficient: (1) practical problems of political action (Downs, 1975); (2) unequal bargaining strengths between prospective employers, and local government (Williamson, 1975); (3) unspecified property rights (Demstz, 1967); (4) the lumpy nature of locational decisions (Isard, 1966); (5) activities of federal job creation programs; and (6) failure of business to create the jobs they promise (HUD, 1982: 85).

LESSONS AND HYPOTHESES

What appears clear is that cities are attempting to protect themselves from the hardships of shifting comparative advantage. The competition among cities to purchase jobs is one protective mechanism, although we know relatively little about how to deal with community economic decline (Heilburn, 1981: 539). However, lessons are emerging. Mazza (1982) developed a handbook, almost a cookbook, describing steps communities can take to deal with potential shutdown. The recipe includes (1) establishing an economic action team, (2) exploring alternative financial and ownership arrangements to maintain the plant, and (3) selecting a recovery strategy (i.e., business attraction, expansion, entrepreneurship, etc.), and (4) identifying financing sources. The Mazza model described steps taken by both Springfield and Fort Wayne. Unfortunately, it worked only for Springfield. Clearly much of a community's economic future remains beyond local control.

The theoretical analysis presented in the previous section implies that the combined welfare of residents of Springfield and Fort Wayne would have been maximized if neither city had offered IH financial incentives. (The lower right quadrant of Figure 14.2 depicts this situation.) It is clear that one of the plants would have remained in operation even if no subsidy was given, especially since the bids were so similar that they offset each other and had no impact on IH's eventual decision. How then did IH get both cities to offer such large assistance packages?

Part of the answer appears to be that the situation was not initially perceived as a zero sum game between the two communities. That is to say, there was an expectation (or at least a hope) that both plants could be saved. At no point did IH explicitly compare the competing offers or say that one city had to match the other's offer. However, the decision was so important to both cities that newspaper accounts and speculations encouraged competition. As negotiations developed and it became clear that one city would lose out to the other, both communities felt intense pressure to present the most attractive package possible. In addition, only representatives of IH knew the size of the subsidies necessary to keep the plant in operation. This information advantage led to greater concessions than would have been offered if local officials had more complete knowledge.

Buss and Redburn (1983) provided a possible explanation of why International Harvester extended the decision until the last moment. They contended that company officials may fear vandalism and reduced productivity if workers are given significant lead time in plant closing notification. By delaying the decision, the workers in both plants were under pressure to be highly productive in the hope of convincing IH to stay.

Another reason for the attractiveness of the two offers can be found in the fact that there were only two cities competing for the IH plant. If the number of cities being considered as a possible site was large, it might not have been worthwhile for all of them to commit resources to developing coalitions of interested parties and to prepare complex assistance packages. (This argument may become less important as the market for jobs develops and "bidding costs" decrease.)

The nature of the jobs at stake was also influential in determining the amount of subsidy offered IH. The threatened jobs were well paying and not easily replaced. These were jobs that generated positive externalities for many local businesses and supported the local economy. Also, local financial institutions had direct interest in the outcome due to the number of mortgages at risk. The fact that IH workers were unionized probably gave them more clout than if they had been employed in the secondary (casual) labor market. Certainly, political considerations were greater in this situation since workers already had the IH jobs. If the subsidy decision had involved a new plant and if the potential employees were not identified or possibly not even current residents, the offers might not have been so generous.

The distressed nature of both cities probably contributed to their desire to maintain the IH plant. Newspaper accounts portrayed bleak prospects for each city if its plant were to close. The bargaining took place at a time when manufacturing activity in the north-central region was in decline and when it was unlikely that alternative uses for the facility or employment for the workers could be found. In addition, the distressed nature of the two cities suggests high external benefits from job maintainance. A growing community with a low rate of unemployment would have enjoyed fewer external benefits and, hence, would have developed a less generous financial package.

The large variety of participants also contributed to the ability of both cities to secure funding. In both cases, the city government was actively involved, but the majority of funds came from other sources. The threat of extensive job loss galvanized both public and private sector interest. The creation of the quasipublic CIC was especially important. Since it was not a public agency, potential political problems were reduced and the sense of private sector partnership enhanced. This may prove to be a useful and enduring contribution to the emerging institutional structure of the market for jobs.

NOTES

1. We wish to thank Richard Wroblewski for his assistance in collecting material for this case.

2. Three plants were to be consolidated into two. However, the third plant, located in Chatham, Ontario, was considered safe from shutdown because of a local content law that requires Canadian production if IH were to have access to the Canadian market.

REFERENCES

BLAIR, J. P., R. FICHTENBAUM, and J. SWANEY (forthcoming) "The market for jobs: locational decisions and the competition for economic development." Urban Affairs Quarterly.

BLUESTONE, B. and B. HARRISON (1982) the Deindustralization of America: Plant Closing, Community Abandonment and the Dismantling of Basic Industry. New York: Basic Books.

BUSS, T. F. and F. S. REDBURN (1983) Shutdown at Youngstown: Public Policy for Mass Unemployment. New York: Albany State University of New York Press.

DEMSTZ, H. (1967) "Toward a theory of property rights." American Economic Review, Papers and Proceedings 57 (May): 347-359.

Department of Housing and Urban Development (1982) An Impact Evaluation of Urban Development Action Grants. Washington, DC: HUD.

DOTSON, A. (1983) "Plant closings and the solution of cities." Journal of American Planning Association 49, 3: 358-362.

DOWNS, A. (1957) An Economic Theory of Democracy. New York: Harper & Row.

GUTHERIE, T. L. (1982) "The cold, hard facts." Fort Wayne Five Year Marketing Plan. Fort Wayne, Indiana.

HEILBRUN, J. (1981) Urban Economics and Public Policy. New York: St. Martin's Press.

ISARD, W. (1966) "Game theory, location theory and industrial agglomeration." Papers and Proceedings of the Regional Science Association, pp. 1-11.

MARSHS, B. and S. SAVILLE (1982) "International Harvester's story: how a great company lost its way." Crain's Chicago Business (November): 19-45.

MAZZA, J. et al. (1982) Shutdown: A Guideline for Communities Facing Plant Closings. Washington, DC: Northeast-Midwest Institute.

Springfield Chamber of Commerce (1982) The Springfield First Aid Kit. Springfield, Ohio.

The Fantus Company (1982) The Advantages of Fort Wayne, Indiana as an Industrial Location.

WILLIAMSON, O. (1975) Markets and Hierarchies. New York: Free Press.

About the Contributors

LYNN W. BACHELOR is Assistant Professor of Political Science at Wayne State University. In addition to her ongoing research on urban economic development politics, she is involved in research on the distribution and resolution of citizen complaints about urban services and on the impact of administrative decentralization on school resource distribution, and she is the author of journal articles on these subjects.

ROBERT L. COOK BENJAMIN is an economist with the U.S. Department of Housing and Urban Development, having previously taught American social and urban history at Yeshiva University. He has authored or coauthored several HUD publications on indicators of urban economic and housing needs, developed the classification of cities used in the 1980 President's Urban Policy Report, and developed the eligibility system for UDAG grants. His most important development, shared with a lovely midwestern wife, is an equally lovely eastern baby girl named Rachel Miriam (1983: October 16).

BERNARD L. BERKOWITZ is President of the Baltimore Economic development Corporation. His previous positions in Baltimore include the Mayor's Physical Development Coordinator and Deputy Director of City Planning. He received his M.A. Degree in City Planning from Columbia University where he also completed a year of graduate studies in economics. He has taught courses in planning and development at The Johns Hopkins University and the Community College of Baltimore. In addition to authoring the Economic Development section of Baltimore City's Comprehensive Plan, he has written numerous articles on economic development for newspapers and journals.

RICHARD D. BINGHAM is Associate Professor of Political Science and Director of the Urban Research Center at the University of

Wisconsin—Milwaukee. He is author or coauthor of books and articles on innovation in local government, state and local finance, and urban economic development. His current research interests are in the areas of fiscal retrenchment and urban management. His latest book, *Researching Decisions in Public Policy and Administration* (Longman, 1982), was edited with Marcus Ethridge.

JOHN P. BLAIR is Chairman of the Department of Economics at Wright State University. He has published articles in a variety of journals and has contributed to previous Urban Affairs Annual Reviews. He is currently studying local job development efforts.

MICHAEL BRINTNALL is a senior analyst with HUD's Office of Policy Development and Research and was director of the Economic Development and Evaluation Division in HUD's Office of Community Planning and Development. He received his Ph. D. in Political Science from MIT.

STUART M. BUTLER, Director of Domestic Policy Studies at the Heritage Foundation, is recognized as the leading authority in the United States on urban "enterprise zones." Butler, a British-born economist, has authored numerous Heritage Foundation studies on "enterprise zones," rent control, nationalized health, and other urban policy matters. Prior to joining Heritage in 1979, Butler was an instructor in economics at Hillsdale College in Michigan for two years. He has been an Honorary Research Fellow for the Institute of United States Studies at the University of London since 1978. Butler was educated at St. Andrews in Scotland, where he received a Bachelor of Science Degree in Physics and Mathematics in 1968, a Master's Degree in Economics in 1971, and a Ph. D. in Economic History in 1978.

GAVIN M. CHEN is currently a senior economist with the Minority Development Agency of the U.S. Department of Commerce. He completed graduate studies in economics and in finance at both the Graduate School of Arts and Sciences and the Graduate School of Business at Washington University. His current research interests are in international economic development of Latin America and the Caribbean. He has authored or coauthored several articles on economic integration, trade, and development. He has also written on entrepreneurship, investment, and industrial structure in the minority business area. He is currently a reader for the *Review of*

Black Political Economy and consults extensively throughout the Caribbean basin.

PAUL R. DOMMEL is the Chairperson and Professor of Political Science at Cleveland State University. He was formerly a Senior Fellow at the Brookings Institution in Washington, D.C., where he directed a longitudinal national study of the Community Development Block Grant Program. He is the author of *The Politics of Revenue Sharing* and coauthor of *Decentralizing Urban Policy*. He has also authored a number of chapters in books and journal articles on urban policy and intergovernmental fiscal relations.

MARCUS E. ETHRIDGE is Assistant Professor of Political Science at the University of Wisconsin—Milwaukee. His published work includes research on the policy effects of legislative oversight, citizen participation procedures, and change in administrative procedures. A continuing interest in the methods and practice of policy analysis culminated in *Reaching Decisions in Public Policy and Administration,* a volume he edited with Richard Bingham. Ethridge is currently involved in a conceptual analysis of the policy implications of several economic models of bureaucratic structure and behavior.

CLAIRE L. FELBINGER is a Ph.D. candidate in Political Science at the University of Wisconsin-Milwaukee. She was formerly a research assistant with the Division of Industrial Science and Technological Innovation of the National Science Foundation. Her research interests are in the areas of intergovernmental policy. Her current research focuses on the adoption of technological innovations in urban areas as a municipal retrenchment strategy.

GARY GAPPERT is a social economist and Director of the Institute for Future Studies in Research at the University of Akron. He is the author of *Post-Affluent America* and coeditor of two previous volumes in the Urban Affairs Annual Reviews series.

PAUL K. GATONS is a senior policy analyst with HUD's Office of Policy Development and Research and was Director of the Urban Studies Division in HUD's Office of Community Planning and Development. Prior to his positions at HUD, he was on the faculties of Georgia State University and Louisiana Technical University and was coauthor of *Economics of Urban Problems: An Introduction* and coeditor of *Economics of Urban Problems: Selected Readings.*

FRANKLIN J. JAMES is Associate Professor in the Graduate School of Public Affairs of the University of Colorado at Denver and Research Director of the University's Institute for Urban and Public Policy Research. He directed HUD's Legislative and Urban Policy Staff during the Carter administration, and has been a senior member of the research staff of the Urban Institute and the Rutgers University Center for Urban Policy Research.

BRYAN D. JONES is Professor and Chairperson of Political Science at Wayne State University. He is the author of books and articles on the subjects of urban service delivery, bureaucracies, and the politics of economic development.

RICHARD V. KNIGHT is an economist specializing in the development of city-regions. He has been working with the National Research Council's Committee on National Urban Policy. His published works include *The Metropolitan Economy* and he is a coeditor of a previous volume in the Urban Affairs Annual Reviews entitled *Cities in the 21st Century.*

LARRY LITVAK is an investment officer in the Asset Management Division of the United States Trust Company, Boston, Massachusetts, responsible for pension fund, endowment, and individual portfolio management. Before joining U.S. Trust, he worked as an advisor to public pension fund sponsors in designing investments with an emphasis on economic development, evaluating portfolio performance, and structuring investment regulations. His clients have included the states of California, Pennsylvania, New York, and New Jersey; the National Governors' Association; and several unions. Mr. Litvak is the author of *Pension Funds and Economic Renewal,* the leading guide on prudent investment of pension funds in economic development. He holds a B.A. in economics from Stanford University and an M.A. in Public Policy from Harvard University.

WILLIAM B. NEENAN is Dean of the College of Arts and Sciences and Professor of Economics at Boston College. His research interests have focused on problems of health economics and urban public finance. Among his published works are *Political Economy of Urban Areas* (Chicago: Markham, 1972) and *Urban Public Economics* (Belmont: Wadsworth, 1981).

ROBERT PREMUS is an economist for the Joint Economic Committee, Congress of the United States. He specializes in regional, industrial policy, and federalism issues for the committee. He received his Ph.D. from Lehigh University and he holds the rank of Professor of Economics at Wright State University. He is on academic leave of absence at this time. His publications include *Location of High Technology Firms and Regional Economic Development* and numerous articles on technological innovation and regional development. Currently he is completing a comprehensive study of venture capital companies in the United States.

BARTON WECHSLER is Assistant Professor of Public Administration at Florida State University. His research is in the areas of public sector strategic management, economic development, and organization behavior.